Reviews of Physiology, Biochemistry and Pharmacology 151

Springer

Berlin
Heidelberg
New York
Hong Kong
London
Milan
Paris
Tokyo

Reviews of

151 Physiology Biochemistry and Pharmacology

Editors
S.G. Amara, Portland • E. Bamberg, Frankfurt
H. Grunicke, Innsbruck • R. Jahn, Göttingen
W.J. Lederer, Baltimore • A. Miyajima, Tokyo
H. Murer, Zürich • S. Offermanns, Heidelberg
G. Schultz, Berlin • M. Schweiger, Berlin

With 15 Figures

Springer

ISSN 0303-4240

ISBN 978-3-642-06043-4 e-ISBN 978-3-540-44423-7

Springer-Verlag is a part of Springer Science+Business Media

springeronline.com

Softcover reprint of the hardcover 1st edition 2004

Editor: Dr. Thomas Mager, Heidelberg
Desk editor: Anne Clauss, Heidelberg
Production editor: Dagmar Orth, Andreas Gösling, Heidelberg
Cover design: design & production GmbH, Heidelberg

Printed on acid-free paper – 14/3150 ag 5 4 3 2 1 0

Rev Physiol Biochem Pharmacol (2004) 151:1–44
DOI 10.1007/s10254-003-0021-1

H. Wegele · L. Müller · J. Buchner

Hsp70 and Hsp90—a relay team for protein folding

Published online: 23 January 2004

Abstract Molecular chaperones are a functionally defined set of proteins which assist the structure formation of proteins in vivo. Without certain protective mechanisms, such as binding nascent polypeptide chains by molecular chaperones, cellular protein concentrations would lead to misfolding and aggregation. In the mammalian system, the molecular chaperones Hsp70 and Hsp90 are involved in the folding and maturation of key regulatory proteins, like steroid hormone receptors, transcription factors, and kinases, some of which are involved in cancer progression. Hsp70 and Hsp90 form a multichaperone complex, in which both are connected by a third protein called Hop. The connection of and the interplay between the two chaperone machineries is of crucial importance for cell viability. This review provides a detailed view of the Hsp70 and Hsp90 machineries, their cofactors and their mode of regulation. It summarizes the current knowledge in the field, including the ATP-dependent regulation of the Hsp70/Hsp90 multichaperone cycle and elucidates the complex interplay and their synergistic interaction.

Abbreviations *TPR:* Tetratricopeptide repeat · *Hsp70:* Heat shock protein 70 · *yHsp70:* Yeast heat shock protein 70 · *Hsp90:* Heat shock protein 90 · *yHsp90:* Yeast heat shock protein 90 · *hHsp90:* Human heat shock protein 90 · *SHR:* Steroid hormone receptor · *GA:* Geldanamycin · *GR:* Glucocorticoid receptor · *PR:* Progesterone receptor · *ER:* Endoplasmic reticulum · *DSG:* 15-Deoxyspergualin · *GR-LBD:* Glucocorticoid receptor ligand binding domain

Introduction

The native three-dimensional structure of proteins, which is of crucial importance for their stability and function, is determined by noncovalent interactions of amino acid side chains.

H. Wegele · L. Müller · J. Buchner (✉)
Institut für Organische Chemie und Biochemie, Technische Universität München, Lichtenbergstrasse 4, 85747 Garching, Germany
e-mail: johannes.buchner@ch.tum.de · Tel.: +49-89-28913341 · Fax: +49-89-28913345

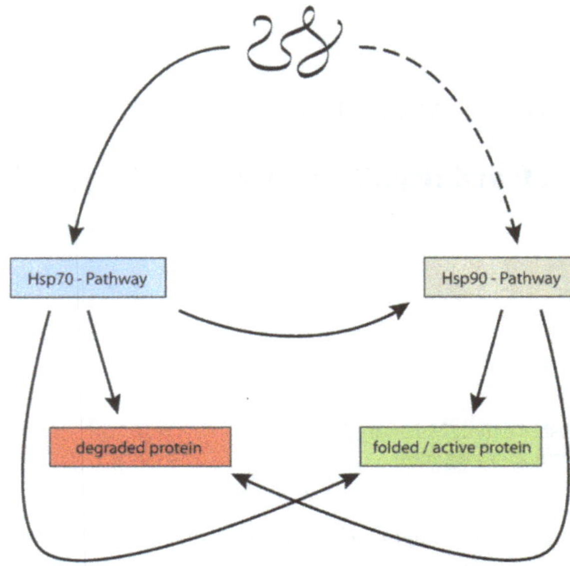

Fig. 1 Protein folding pathways. Two major pathways for an unfolded or partially folded polypeptide chain are depicted. While the direct interaction of Hsp70 with unfolded/nascent proteins and the transfer of some of these to Hsp90 is well established, it is unclear to what extent Hsp90 interacts directly with nonnative proteins

The process of structure acquisition, termed protein folding, is a spontaneous process. In the cell, however, some proteins do not achieve or maintain their correct structure without the assistance of helper proteins, the so-called molecular chaperones. In eukaryotes, several different chaperones determine the fate of partially or completely unfolded proteins. In this review, we focus on two major chaperone classes and their interplay (Fig. 1).

One of the most frequently used folding systems in the cell is the Hsp70 machinery, which is responsible for the correct folding of a wide variety of protein substrates. However, some proteins are processed by Hsp70 and then transferred to the Hsp90 machinery. There is some evidence that Hsp90 is also able to act independently of Hsp70. Each chaperone folding pathway can either lead to folded, functional proteins or to degradation.

Compared to other chaperones, a wide variety of in vivo substrates is known for the Hsp70/Hsp90 chaperone network (Table 1) (Buchner 1999; Jolly and Morimoto 2000; Mayer et al. 2001; Picard 2002; Pratt and Toft 1997; Richter and Buchner 2001; Smith and Toft 1993; Walter and Buchner 2002; Young et al. 2001; Zylicz et al. 2001). In many cases, these proteins have been shown by coimmunoprecipitation to exist in cytosolic heterocomplexes with Hsp90. More recently, ansamycin antibiotics, such as geldanamycin, and radicicol, which inhibit Hsp90 function (Whitesell et al. 1994; see the section entitled "Modulation of the Hsp90 ATPase") provided a rapid means of screening for Hsp90-regulated targets. In the last five years, the number of established substrates or "client proteins" that are regulated by Hsp90 has tripled. These substrates include transcription factors such as steroid hormone receptors and some proto-oncogenic serine/threonine and tyrosine kinases like Raf and Src in higher eukaryotes (for reviews see Buchner 1999; Picard 2002; Pratt and Toft 2003).

The Hsp90 system is a complicated machinery. The seemingly close collaboration with Hsp70 and the large number of cofactors makes it unique. To understand the underlying functional properties, it is important to realize the properties of the two major components and the evidence for their synergistic interaction in vivo and in vitro.

Table 1 Hsp90 client proteins. Summary of known Hsp90 client proteins (taken, extended and modified from Pratt and Toft 2003 and Picard 2002)

	Protein	Reference
Transcription	Androgen receptor	Joab et al. 1984; Veldscholte et al. 1992
	Aryl hydrocarbon (Ah) receptor	Denis et al. 1988; Perdew 1988
	Ecdysone receptor	Arbeitman and Hogness 2000
	Estrogen receptor	Joab et al. 1984; Sabbah et al. 1989
	Glucocorticoid receptor	Ben Or 1989
	Heme activator protein (Hap1)	Zhang et al. 1998
	HSF-1	Ali et al. 1998
	Hypoxia-inducible factor-1α	Minet et al. 1999
	Mineralocorticoid receptor	Rafestin-Oblin et al. 1989
	MTG8 myeloid leukemia protein	Komori et al. 1999
	p53	Blagosklonny et al. 1996
	Progesterone receptor	Catelli et al. 1985; Schuh et al. 1985
	Retinoid receptor	Holley and Yamamoto 1995
	Sim	McGuire et al. 1995
	SV40 large T antigen	Miyata and Yahara 2000
	Tumor promotor-specific binding protein	Hashimoto and Shudo 1991
	v-erbA	Privalsky 1991
Polymerases	Telomerase	Holt et al. 1999
	Hepatitis B virus reverse transcriptase	Hu and Seeger 1996
	DNA-polymerase α	Crevel et al. 2001
Kinases	3-Phosphoinositide-dependent kinase-1	Fujita et al. 2002
	Akt	Sato et al. 2000
	Bcr-Abl	An et al. 2000
	Calmodulin-regulated eEF-2 kinase	Palmquist et al. 1994
	Casein kinase II	Miyata and Yahara 1992
	Cdc2	Munoz and Jimenez 1999
	Cdk4	Dai et al. 1996; Stepanova et al. 1996
	Cdk6	Mahony et al. 1998
	Cdk9	O'Keeffe et al. 2000
	c-Mos	Fisher et al. 2000
	Epidermal growth factor receptor	Stancato et al. 1997
	ErbB2	Xu et al. 2001
	Focal adhesion kinase	Masson-Gadais et al. 2003
	Hck	Scholz et al. 2000
	Heme-regulated eIF-2α kinase	Matts and Hurst 1989; Rose et al. 1987
	Insulin receptor	Takata et al. 1997
	Insulin-like growth factor receptor	Jerome et al. 1991
	Kinase suppressor of ras (KSR)	Stewart et al. 1999
	Lymphoid cell kinase $p56^{lck}$	Hartson and Matts 1994
	MAK-related kinase	Miyata et al. 2001
	Male germ cell-associated kinase MAK	Miyata et al. 2001
	MEK (MAP kinase kinase)	Stancato et al. 1997
	Mik1	Goes and Martin 2001
	Mitogen-activated protein kinase MOK	Miyata et al. 2001
	Phosphatidylinositol 4-kinase	Flanagan and Thorner 1992
	Pim-1	Mizuno et al. 2001
	PKR	Donze et al. 2001
	Platelet-derived growth factor receptor	Sakagami et al. 1999
	Polo mitotic kinase	de Carcer et al. 2001
	Raf family kinases: v-Raf, c-Raf, B-Raf, Gag-Mil, Ste11	Jaiswal et al. 1996; Louvion et al. 1998; Lovric et al. 1994; Stancato et al. 1993; Wartmann and Davis 1994
	Receptor-interacting protein (RIP)	Lewis et al. 2000
	Sevenless PTK	Cutforth and Rubin 1994
	Swe1	Goes and Martin 2001

Table 1 (continued)

	Protein	Reference
Kinases	Translation initiation factor kinase Gcn2	Donze and Picard 1999
	Tropomyosin related kinase B (trkB)	Bernstein et al. 2001
	v-fes	Lipsich et al. 1982
	v-fps	Lipsich et al. 1982
	v-fgr, c-fgr	Hartson and Matts 1994; Ziemiecki et al. 1986
	v-Src, c-Src	Brugge et al. 1981; Hutchison et al. 1992; Oppermann et al. 1981
	v-yes	Lipsich et al. 1982
	Wee1	Aligue et al. 1994
Others	Actin	Koyasu et al. 1986
	Apaf-1	Saleh et al. 2000
	Apoprotein B	Gusarova et al. 2001
	Atrial natriuretic peptide receptor	Kumar et al. 2001
	Calponin	Bogatcheva et al. 1999
	Centrin/centrosome	Uzawa et al. 1995
	Cna2 (catalytic subunit of calcineurin)	Imai and Yahara 2000
	CFTR	Loo et al. 1998
	Endothelial NOS	Garcia-Cardena et al. 1998
	Fanconi anemia group C protein	Hoshino et al. 1998
	G protein $\beta\gamma$	Inanobe et al. 1994
	Gα0	Busconi et al. 2000
	Gα12	Vaiskunaite et al. 2001
	Guanylate cyclase (β-subunit)	Venema et al. 2003
	HETE binding complex	Herbertsson et al. 1999
	Inducible NOS	Joly et al. 1997
	Lysosomal membrane	Agarraberes and Dice 2001
	Macrophage scavenger receptor	Nakamura et al. 2002
	Aminoacyl t-RNA synthetase	Kang et al. 2000
	Mdm2	Peng et al. 2001
	Myosin	Kellermayer and Csermely 1995
	Neuronal NOS	Bender et al. 1999
	Pancreatic bile salt-dependent lipase	Bruneau et al. 1998
	Erythrocyte membrane protein (Plasmodium falciparum)	Banumathy et al. 2002
	Protease-activated receptor 1 (PAR-1)	Pai et al. 2001
	Proteasome	Tsubuki et al. 1994
	Reovirus protein σ1	Gilmore et al. 1998
	β-Galactosidase M15 truncation mutant	Abbas-Terki and Picard 1999
	Thyroglobulin	Muresan and Arvan 1997
	Tubulin	Sanchez et al. 1988
	Unassembled immunoglobulin chains	Melnick et al. 1992

The Hsp70/Hsp90 chaperone cycle—an overview

In 1966, Toft and Gorski (Toft and Gorski 1966) demonstrated that estradiol added to the cytosol of rat uteri was incorporated in a large 9S complex. Subsequently, ultracentrifugation experiments indicated that ligand-free steroid hormone receptors (SHRs) are part of large multiprotein complexes (Pratt and Toft 1997; Wilson et al. 1977). This represents the first direct demonstration of hormone binding to a receptor protein in a cell-free system. SHRs are soluble cytosolic proteins composed of three domains: a hormone binding

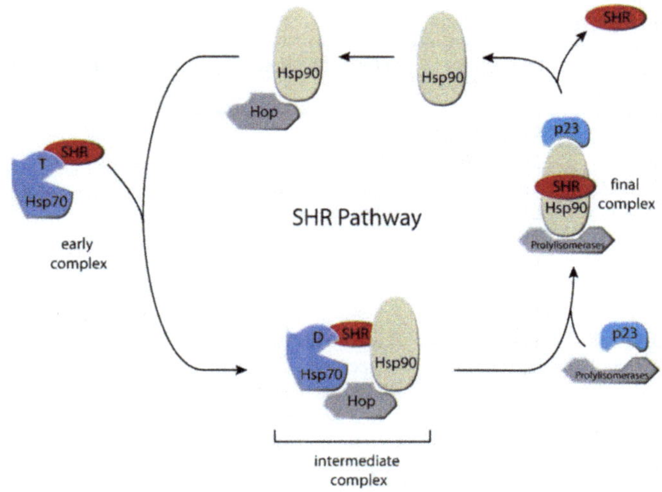

Fig. 2 Basic chaperone cycle for activation of steroid hormone receptors. Hsp70 delivers the inactive steroid hormone receptor (*SHR*) to the "intermediate chaperone complex," consisting of Hsp70, Hsp90, and the scaffold protein Hop. Hop and Hsp70 are exchanged for a prolylisomerase and p23 to yield the "final complex." Binding of the hormone to the now activated receptor prevents the receptor from rebinding to the complex. *T* ATP-bound form, *D* ADP-bound form

domain, a DNA binding domain, and a transactivator domain which is responsible for the activation of the genes to be regulated (Evans and Hollenberg 1988). The binding of the hormone promotes the transformation from the inactive to the active DNA binding form. Coimmunoprecipitation experiments enabled the identification of Hsp90 as an additional component (Sanchez et al. 1987; Ziemiecki et al. 1986). The stable association of a heat shock protein with a SHR was completely unexpected at that time. Interestingly, it was not possible to reconstitute the 9S complex using only SHR and Hsp90, suggesting that complex formation is not a simple binding reaction (Scherrer et al. 1990). However, assembly of in vitro translated receptor into active multiprotein complexes was possible in rabbit reticulocyte lysate (Denis and Gustafsson 1989), indicating that additional factors are involved in the assembly process (Dalman et al. 1989; Scherrer et al. 1990; Smith et al. 1990). Two of them were identified as Hsp70 and Hop (Perdew and Whitelaw 1991; Smith et al. 1993). Hsp70 was already known to be a general molecular chaperone (Beckmann et al. 1990; Schroder et al. 1993; Tilly et al. 1983), whereas Hop was shown later to bind to both Hsp70 and Hsp90 (Hernandez et al. 2002b; Perdew and Whitelaw 1991; Scheufler et al. 2000; Smith et al. 1993; Wegele et al. 2003a). Hop does not act as a chaperone itself (Bose et al. 1996; Freeman et al. 1996).

In 1993, David Smith proposed a chaperone cycle for the activation of SHRs (Smith 1993). Based mainly on results of immunoprecipitation experiments, he suggested that there is an ordered sequence of association and dissociation of cofactors to and from the Hsp90-SHR complex which is required for the activation of the target protein. The discovery of additional components led him to expand his original suggestion of the chaperone cycle (Smith et al. 1995; Fig. 2).

This bold suggestion has been important for our understanding and further investigation of the Hsp90 chaperone system.

Table 2 Hsp90 and Hsp70 co-factors. Summary of known Hsp70 and Hsp90 co-factors, their yeast homologues and their cellular function

Human	Yeast homologue	Interaction	Cellular function	Essential (yeast)	Hsp70 co-factor	Hsp90 co-factor
Hsp90	yHsp90	Via Hop/Sti1	Chaperone	+	–	–
Hsp70	Ssa1–4	Via Hop/Sti1	Chaperone	+	–	–
Hop	Sti1	TPR	Complex assembly	–	+	+
p23	Sba1	Direct	Substrate-release/ complex assembly	–	–	+
TTC4	Cns1	TPR	Function unknown	+	+	+
FKBP52	–	TPR	Prolylisomerase	n.d.	–	+
FKBP51	–	TPR	Prolylisomerase	n.d.	–	+
Cyp40	Cpr6, Cpr7	TPR	Prolylisomerase	–	–	+
Hdj1	Ydj1, Sis1, Zuo1	Direct	Chaperone, ATPase activator	+	+	–
Bag1	Snl1, Mge1	Direct	Nucleotide release	+	+	–
CHIP	–	TPR	Ubiquitin ligase	–	+	+
PP5	Ppt1	TPR	Phosphatase	–	–	+
XAP2	–	TPR	Prolylisomerase-related protein	n.d.	–	+
Cdc37/p50	Cdc37	Direct	Signal transduction	+	–	+
Aha1	Aha1/Hch1	Direct	Hsp90 ATPase regulator	–	–	+

In the first step of this cycle, the free glucocorticoid receptor (GR) with a closed binding site for steroid hormones is bound by Hsp70. The binding of GR involves ATP hydrolysis by Hsp70.

The substrate protein bound to Hsp70 in the intermediate complex is brought into contact with Hsp90 via Hop. Hop and Hsp70 as parts of the "intermediate complex" are thought to be exchanged for a prolylisomerase/immunophilin and p23 to yield the "final complex." In the maturation from the "intermediate complex" to the "final complex," the substrate has to be transferred from Hsp70 to Hsp90. A complex consisting of Hsp70, Hop and Hsp90 was shown to be sufficient to transform SHRs from the nonsteroid binding form to the steroid binding form in vitro (Dittmar and Pratt 1997). Further, it was demonstrated that steroid binding to the SHR in this complex was not very efficient unless another protein called p23 was present (Hutchison et al. 1995; Johnson et al. 1994; Johnson and Toft 1994). The interaction between the Hsp90 heterocomplex and its substrates seemed to be stabilized by p23 (Dittmar and Pratt 1997).

Binding of the hormone to the SHR leads to receptor dimerization and appears to decrease its affinity for Hsp90, making the reloading of Hsp90 with the same receptor via Hsp70 and Hop impossible (Smith and Toft 1993). It was shown that there are preexisting complexes composed of Hsp70, Hsp90, and Hop even in the absence of receptor in the cell (Smith et al. 1992).

This system is not restricted to higher eukaryotes and SHR activation. In *Saccharomyces cerevisiae*, a similar multichaperone complex consisting of an Hsp70 component (the Ssa proteins), Sti1 (the yeast homologue of Hop), and yeast Hsp90 (yHsp90) was found (Chang and Lindquist 1994; Wegele et al. 2003a). Successful heterologous expression of the substrates glucocorticoid receptor and v-Src in yeast (Nathan and Lindquist 1995) shows the conservation of the chaperone mechanism from yeast to man. Genetic studies in yeast and fruit fly revealed that Hsp90 is an essential protein (Arbeitman and Hogness 2000; Chang et al. 1997; Chang and Lindquist 1994; Duina et al. 1996).

Most interestingly, this complex network consisting of Hsp90 and Hsp70 and numerous cofactors (Table 2) does not seem to exist in the prokaryotic and archaeal domains of life. However, in contrast to archaea, in many prokaryotic organisms, Hsp70 and Hsp90 are present.

The Hsp70 chaperone system

The Hsp70 proteins constitute the central part of an ubiquitous chaperone system that is present in eukaryotic cells, in eubacteria, and in many archaea (Macario et al. 1999).

In *E. coli*, at least three cytosolic members of the Hsp70 family exist (Itoh et al. 1999), with DnaK as the most prominent member. In eukaryotes, Hsp70 proteins were found in the cytosol, in chloroplasts, mitochondria, and the lumen of the endoplasmic reticulum (ER). In archaea, Hsp70 is present in halophiles and in some, but not all methanogens, e.g., it is missing in the extreme thermophile *Methanococcus jannaschii* (Leroux 2001; Macario et al. 1999).

In mammals, two isoforms exist in the cytoplasm, a 73-kDa form that is constitutively expressed (Hsc70) and a 72-kDa, stress-inducible form (Hsp70). This highlights the general principle that some Hsp70s fulfill important functions under physiological conditions and others, which are inducible, are important under stress conditions. Recent studies also postulate extracellular localized Hsp70, which may play a key role in the induction of the cellular immune response (Milani et al. 2002; Multhoff 2002).

It has been shown that Hsp70 binds promiscuously to a wide variety of newly synthesized or unfolded proteins (Beckmann et al. 1990; Frydman et al. 1994; Hendrick and Hartl 1993; Mogk et al. 1999). Binding occurs regardless of secondary and tertiary structure by recognition of hydrophobic sequences. For eukaryotic, cytosolic Hsc70 a direct interaction with nascent, ribosome-associated polypeptide chains was observed (Beckmann et al. 1990; Hansen et al. 1994; Kim et al. 1998; Nelson et al. 1992).

The importance of Hsp70 proteins is highlighted by the fact that in addition to the general assistance of protein folding, they are involved in important cellular processes, like the transport of proteins across membranes (Herrmann and Neupert 2000; Pfanner 2000; Schatz and Dobberstein 1996), the disassembly of clathrin-coated vesicles (Schmid et al. 1985; Schmid and Rothman 1985), and the regulation of the heat shock response (Abravaya et al. 1992; Yura et al. 1993), among others.

The common function of Hsp70 in these processes appears to be the binding of short hydrophobic segments in partially folded polypeptides, thereby preventing aggregation and allowing correct folding. The broad spectrum of cellular functions of Hsp70 proteins is achieved through the following strategies. First, the amplification and diversification of Hsp70 genes in evolution has generated functionally specialized Hsp70 chaperones. For example, the extensively studied Hsp70 family in *Saccharomyces cerevisiae* comprises 14 members (Werner-Washburne and Craig 1989). Nine of the 14 yeast Hsp70 proteins are located in the cytosol and can be divided into the subfamilies Ssa, Ssb, Sse, and Ssz (Ss denotes stress seventy-related; a,b,e, and z indicate the subfamily). The Ssa subfamily comprises four members (Ssa1–Ssa4). The Ssa proteins, which are most closely related to the major cytosolic Hsp70s of mammalian cells, are involved in translocation of proteins across membranes and in the regulation of the heat shock response (Deshaies et al. 1988). Also located in the cytosol are Ssb1 and Ssb2, which differ from each other by only four amino acids and from the members of the Ssa family by 37%. They seem to have a more specialized function as they bind to translating ribosomes and can be crosslinked to the nascent chains (Nelson et al. 1992; Pfund et al. 1998). This association, in addition to the fact that strains lacking Ssb are hypersensitive to certain inhibitors of protein synthesis, suggests that this class of Hsp70s may be involved in translation and/or very early folding events on the ribosome. Another example showing the functional diversity of Hsp70s in the same compartment are Ssa1 and Ssa2, which are about 97% identical in sequence. It was shown that overexpression of Ssa1, but not Ssa2, cured cells of [URE3], which is the prion form of the yeast protein Ure2 (Schwimmer and Masison 2002).

A second strategy for functional differentiation is a large set of cochaperones which regulate the activity of the Hsp70 chaperones. In baker's yeast, the Ssa1 ATPase is regulated by the Hsp40 cochaperones Ydj1 and Sis1 (Cyr et al. 1992; Lu and Cyr 1998), whereas another cytosolic isoform of Hsp70, the Ssz1 protein, has been found in complex with the Hsp40 protein Zuo1 (Gautschi et al. 2001; Hundley et al. 2002). This Ssz1-Zuo1 complex, also called ribosome-associated complex (RAC), acts on nascent chains while they are being synthesized (Gautschi et al. 2001; Gautschi et al. 2002; Hundley et al. 2002). The genome sequence of *Saccharomyces cerevisiae* suggests that 14 different Hsp40 proteins reside in the yeast cytosol. The selective recruitment of these cochaperones allows further fine-tuning of the basic Hsp70 activity.

Third, Hsp70 homologues cooperate with other chaperone systems to broaden their activity spectrum. In the yeast system, it was shown that specifically the members of the Ssa class are part of a Hsp70/Hsp90 multichaperone complex in *Saccharomyces cerevisiae* (Chang and Lindquist 1994; Wegele et al. 2003a) which allows the transfer of certain substrates from the Hsp70 to the Hsp90 system.

Structural and functional properties of Hsp70

The substrates of Hsp70 proteins are unrelated in sequence and structure and represent a large spectrum of folding conformers ranging from completely unfolded nascent chains emerging at ribosomes and translocation pores (Gautschi et al. 2002; Hundley et al. 2002) to native regulatory proteins such as clathrin (Schmid et al. 1985), transcription factors (including σ^{32}, HSF, steroid hormone receptors; Abravaya et al. 1992; Gamer et al. 1992; Pratt et al. 1992), kinases (including Raf, eIF2α-kinase; Chang et al. 1994; Song et al. 2001), DNA replication proteins (including λP, RepE, RepA; Wickner et al. 1991) and p53 (Hupp et al. 1992). In addition, Hsp70 interacts indiscriminately with nonnative polypeptides, such as a broad spectrum of heat-denatured proteins of *E. coli* extracts (Mogk et al. 1999).

Besides Hsp70 proteins with a general chaperone function and a wide substrate spectrum, other Hsp70 homologues such as *E. coli* HscA (Hsc66) and yeast mitochondrial Ssq1, which interact with iron-sulfur cluster-containing proteins (Hoff et al. 2000; Lutz et al. 2001; Voisine et al. 2001), are highly specialized. The differences between these two classes of Hsp70 proteins are poorly understood, but may include the selective recruitment of Hsp40/DnaJ cochaperones and changes in the substrate-binding groove.

The broad substrate specificity of "general" Hsp70s implies a rather degenerative binding motif. It was shown by NMR and X-ray crystallography that DnaK binds its substrate in an extended conformation. Peptide library screening for substrates of DnaK and the ER-residing Hsp70 BiP revealed that these mainly contain hydrophobic residues such as leucine, tyrosine, and tryptophan (Blond-Elguindi et al. 1993a; Flynn et al. 1991; Gragerov et al. 1994; Knarr et al. 1995, 1999; Landry et al. 1992; Rudiger et al. 1997a, b). A typical binding site for DnaK consists of a core of five consecutive residues enriched in hydrophobic amino acids and flanking regions enriched in basic amino acids. Such binding sequences are recognized with affinities of 5 nM–5 μM (Bukau and Horwich 1998) and occur frequently within protein sequences, e.g., approximately one DnaK binding site for every 36 residues (Rudiger et al. 1997b). They are typically buried within the hydrophobic core of the folded proteins or in domain contact sites. In addition, the association of Hsp70 proteins with native proteins seems to require exposed loop regions on the surface of the target proteins (Fink 1997; Greene et al. 1995).

Hsp70s are generally composed of a 44-kDa amino-terminal ATPase domain, a 18-kDa peptide binding domain and a 10-kDa carboxy-terminal variable domain of unknown function (Flaherty et al. 1990; Zhu et al. 1996). The carboxy-terminal domains are the most variable in sequence between the different Hsp70s. Some nonorganellar eukaryotic Hsp70s terminate with a conserved EEVD sequence whose presence is important for the binding of cofactors.

So far it has not been possible to crystallize a full length member of the Hsp70 family, suggesting conformational flexibility in the domain arrangement.

The X-ray crystallographic structure of the DnaK peptide binding domain (residues 389–607) has revealed a novel tertiary fold (Zhu et al. 1996). The structure was solved in complex with a seven residue peptide (Asn-Arg-Leu-Leu-Leu-Thr-Gly) (Fig. 3A).

The peptide binding domain consists of two structural units, an 8-stranded antiparallel β-sandwich with a hydrophobic groove on its upper side, and an α-helical domain that sits on top of it. Loops, that connect the β-sheets, form a hydrophobic groove of 5×7 Å. This is the binding pocket, where the central leucines of the bound peptide are buried. Although biochemical studies suggested the requirement of at least seven residues for optimal peptide affinity (Flynn et al. 1991), only the central five residues of the peptide show substantial interactions with the peptide binding domain in the crystal. Van der Waals contacts between side chains of Hsp70 and side chains of the unfolded polypeptide are responsible for the preference of the peptide binding groove for hydrophobic peptides.

A long α-helix covers this groove, but does not actually contact the peptide. It has been suggested that this helix acts as a "lid" to stabilize peptide complexes and to control the kinetics of peptide binding and release (Mayer et al. 2000). The helix is bound to the peptide binding groove through the polypeptide backbone at the amino-terminus and by salt bridges at the carboxy-terminus. Disruption of one of the salt bridges destabilizes peptide binding (Ha et al. 1997), supporting the hypothesis that the helix serves as a lid, controlling access of the peptide to the groove. Under circumstances where the helix is closed over the groove, peptide binding and release would be relatively slow; however, rapid binding and release was observed if the helix swung outward to allow unhindered access to the groove (Pierpaoli et al. 1998; Rudiger et al. 2000).

Hsp70s have the ability to associate to dimers and other low-order oligomers. Dimerization mimics the interactions of Hsp70 proteins with target polypeptides and denatured proteins in many regards. It can be reversed by the addition of short peptides (suggesting competition for the same peptide binding site) or by addition of ATP (Benaroudj et al. 1995; Blond-Elguindi et al. 1993b; Knarr et al. 2002). The region of Hsp70 proteins responsible for intermolecular association is not known with certainty. Experiments with a Hsc70 construct lacking a carboxy-terminal region showed that this construct is substan-

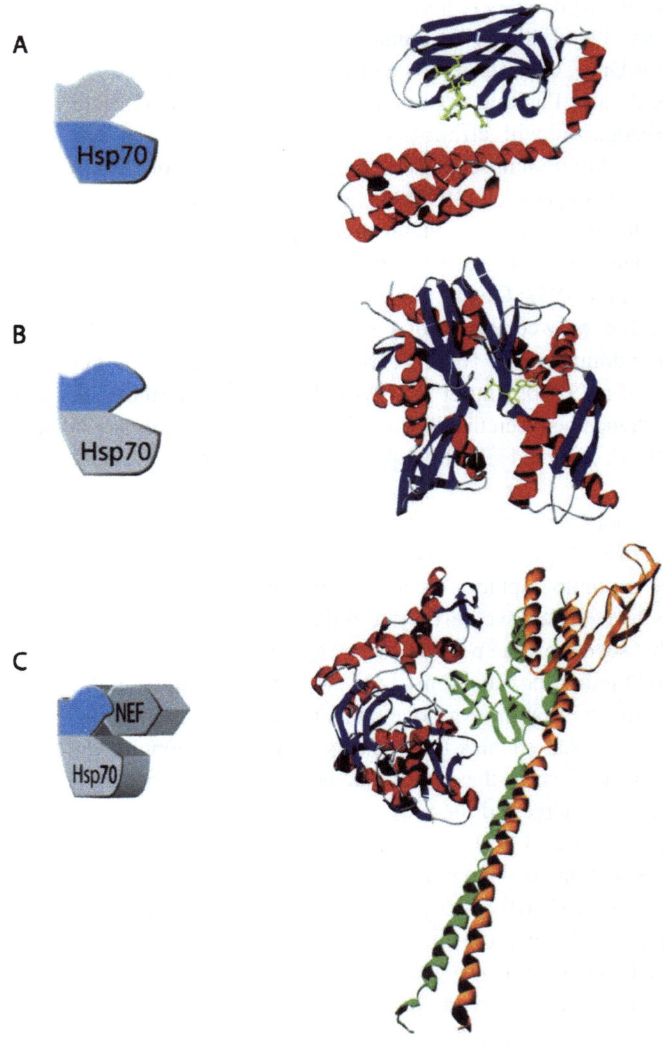

Fig. 3A–C Crystal structure of Hsp70. **A** Substrate binding domain of Hsp70. NMR structure of the 25-kDa *E. coli* DnaK substrate binding domain (PDB entry 1DKX).The bound peptide is depicted in *green*. **B** ATPase domain of Hsp70. Crystal structure of the 42-kDa amino-terminal ATPase domain of human Hsp70 (PDB entry 1HJO). The bound nucleotide ADP is depicted in *green*. **C** Binding of GrpE to Hsp70. Crystal structure of a fragment of the nucleotide exchange factor GrpE (*green/orange*) bound to the ATPase domain of the molecular chaperone DnaK (*red/blue*) (PDB entry 1DKG)

tially less prone to intermolecular association (Wilbanks et al. 1995). This led to the idea that the glycine-rich carboxy terminal 30–50 amino acid residues may be involved.

Binding of ATP to Hsp70 seems to promote conformational rearrangements in the entire molecule, which results in changes in the affinity for peptide substrates (Ha and McKay 1995; Theyssen et al. 1996).

The ATPase domain of Hsp70 is comprised of two subdomains separated by a cleft that contains the nucleotide binding site (Flaherty et al. 1990) (Fig. 3B). The nucleotide is bound at the base of the cleft together with one Mg^{2+} and two K^+ ions and is almost entirely inaccessible to the solvent. The first available structure of an Hsp70 ATPase domain clearly showed a striking similarity to that of actin (Flaherty et al. 1991; Kabsch et al. 1990). Both have an identical folding topology and a tertiary structure similar to the equivalent regions in hexokinase (Fletterick et al. 1975) and glycerol kinase (Hurley et al. 1993). In the meantime, a number of Hsp70 ATPase structures have been determined and the various structures are highly conserved (Harrison et al. 1997; Osipiuk et al. 1999).

The Hsp70 ATPase activity

Hsp70 proteins are involved in the folding of polypeptide chains in an ATP-dependent manner. Cycles of substrate binding and release are coupled to ATP-binding, hydrolysis, and nucleotide exchange (Banecki and Zylicz 1996; McCarty et al. 1995; Palleros et al. 1993). In its ATP-bound form, eukaryotic Hsp70 has a low affinity for substrate proteins, whereas in the ADP-bound form, it exhibits a high substrate binding affinity. Therefore, ATP-hydrolysis converts Hsp70 from the "fast-binding, fast-release" to the "slow-binding, slow-release" form. This locks in the substrate and stabilizes its interaction with Hsp70 (Greene et al. 1995; McCarty et al. 1995; Palleros et al. 1993; Schmid et al. 1994). The general consensus that emerges for several members of the Hsp70 family is that the steady state ATPase turnover rate is relatively slow. The reported k_{cat} values vary between 0.02 min^{-1} and 1.0 min^{-1} with the faster rates generally measured under conditions of peptide stimulation of the ATPase activity (Ha et al. 1999). The steady state K_m values for MgATP are between 0.1 and 1 µM.

Within the enzymatic cycle, ATP hydrolysis is relatively slow and, under many conditions, it appears to be the rate-limiting step of the cycle (Gao et al. 1993; McCarty et al. 1995; Theyssen et al. 1996; Wegele et al. 2003a). In contrast, P_i release is rate limiting for the isolated ATPase domain of Hsc70 (Ha and McKay 1994).

It had been reported previously that the ATPase activity of the amino-terminal domain of mammalian cytosolic Hsc70 is several-fold higher than that of the intact protein, but it is similar to the peptide-stimulated activity (O'Brien and McKay 1993). This decrease in activity of the ATPase domain in the context of the intact protein suggests that the carboxy-terminal domain, in the absence of bound peptide, represses the activity of the ATPase domain. This is also true for other Hsp70 proteins like Ssa1 and the Ssb proteins in *Saccharomyces cerevisiae* (Lopez-Buesa et al. 1998).

The ATPase activity requires Mg^{2+} and is dependent on monovalent ions like K^+ (O'Brien and McKay 1995). The affinity of Hsp70 for both MgATP and MgADP is submicromolar, with reported values ranging from 0.01 to 1 µM (Gao et al. 1994; Ha and McKay 1994; Klostermeier et al. 1998; Theyssen et al. 1996). In the absence of monovalent ions, the affinities for MgATP are weaker (1–10 µM). Significantly smaller (e.g., Na^+ or Li^+) or larger (e.g., Cs^+) monovalent ions give a substantially lower ATPase activity (O'Brien and McKay 1995). While Mg^{2+} complexes the phosphate groups of ATP, crystallographic studies have identified two specific monovalent ion binding sites in the nucleotide binding cleft of the ATPase domain. Binding of these ions influences the exact geometry and electrostatic interactions of the protein-metal-nucleotide complex (Wilbanks and McKay 1995).

In the chemical hydrolysis pathway, MgATP initially binds in a conformation that is not correctly aligned for hydrolysis, possibly with an H_2O molecule intercalated between the Mg^{2+} ion and the γ-phosphate of the nucleotide (Ha et al. 1999). Efficient catalysis requires the correct positioning of the Mg^{2+} ion, a task that needs an electrostatic surrounding which may be established by the binding of two K^+ ions near the Mg^{2+} ion. Then, the γ-phosphate rearranges to form a "tight" β, γ-bidentate complex with the Mg^{2+} ion. This is proposed to align the terminal phosphate of the nucleotide for in-line attack by an H_2O molecule or OH^- ion. A model of a hypothetical transition state of the hydrolysis reaction suggested an experimentally observed H_2O molecule that is hydrogen bonded to the ε-amino group of the side chain of Lys71 in the Hsc70 ATPase fragment as a candidate for a catalytic H_2O molecule (Ha et al. 1999). Lys71 could then act either to accept a proton from the H_2O molecule or to stabilize an OH^- ion. Subsequent studies have shown that mutagenesis of Lys71 completely abolishes ATPase activity (O'Brien et al. 1996). After ATP hydrolysis, P_i is released and ADP is exchanged for ATP to allow the cycle to begin anew.

Modulation of the Hsp70 ATPase activity

To meet the needs of the dynamics of protein synthesis and structure maintenance, the activity of the Hsp70 chaperones is stringently regulated by cofactors which affect critical steps of the ATPase cycle (Bukau and Horwich 1998; Cyr et al. 1992; Lu and Cyr 1998). These effector proteins accelerate the low ATPase activity of Hsp70 by stimulating the hydrolysis step or by facilitating nucleotide exchange. A universal and diverse class of activators of the hydrolysis step are the DnaJ/Hsp40 proteins. They consist of multiple functional domains. One of the domains, the ~75-amino acid J domain, is conserved in all Hsp40 chaperones (Banecki and Zylicz 1996; Cheetham and Caplan 1998; Liberek et al. 1991; McCarty et al. 1995). Mutational analysis revealed that this domain is essential for the interaction between Hsp40 and Hsp70 (Cyr and Douglas 1994). Interestingly, Hsp40s also interact with unfolded polypeptides and may transfer them to Hsp70.

A current model of how Hsp40 and Hsp70 cooperate assumes that a polypeptide substrate first binds to Hsp40 (Hernandez et al. 2002a; Schroder et al. 1993). The substrate binding motif of DnaJ, an *E. coli* Hsp40, consists of a hydrophobic core of approximately eight residues enriched in aromatic, large aliphatic hydrophobic residues and arginine (Rudiger et al. 2001). The Hsp40-protein complex then associates with Hsp70-ATP, the protein substrate is transferred to the open peptide-binding cleft of Hsp70, and locked in by the subsequent Hsp40-stimulated hydrolysis of ATP. The functional cycle of Hsp70 is completed by the exchange of ADP for ATP, which shifts the chaperone back into its dynamic mode and allows the bound polypeptide to dissociate. In the presence of Hsp40, ATP hydrolysis is accelerated up to tenfold (Cheetham and Caplan 1998; Liberek et al. 1991) and may no longer be the rate-limiting step. In this case, the second potentially slow reaction of the cycle, nucleotide exchange, can also be stimulated. In *E. coli*, the cochaperone responsible for this is GrpE (Liberek et al. 1991; Packschies et al. 1997). GrpE binds to the ADP-bound and nucleotide-free states of DnaK with high affinity (K_D=1 nM; Schonfeld et al. 1995) (Fig. 3C). Addition of ATP leads to the dissociation of the complex (Schonfeld et al. 1995; Zylicz et al. 1987). The superposition of the structure of the nucleotide-free ATPase domain of DnaK in complex with GrpE (Harrison et al. 1997) with the structure of the ADP- and P_i-bound ATPase domain of Hsc70 (Flaherty et al. 1990) shows that a subdomain in DnaK is rotated outwards, thereby opening up the nucleotide binding cleft. The mechanism by which GrpE accelerates the nucleotide release thus seems to rely on an active opening of the nucleotide binding cleft. Furthermore, GrpE stabilizes the open conformation of the nucleotide binding pocket, which facilitates the rapid binding of ATP to the nucleotide-free state of DnaK (Brehmer et al. 2001; Packschies et al. 1997).

This model relies mainly on data for the *E. coli* system. In eukaryotes, numerous homologues of DnaJ have been identified in the cytoplasm and in organelles. In yeast, for example, the cochaperones Ydj-1 and Sis1, which belong to the Hsp40 family, stimulate the ATPase activity of Ssa1 approximately tenfold (Cyr et al. 1992; Cyr and Douglas 1994; Lu and Cyr 1998).

In contrast, the search for GrpE homologues in eukaryotes was so far only successful in mitochondria (Bolliger et al. 1994; Laloraya et al. 1994; Nakai et al. 1994). The mitochondrial GrpE homologue in yeast, Mge1, cooperates with mitochondrial Hsp70 in the import of proteins from the cytoplasm (Dekker and Pfanner 1997). Although not sequence-related, the eukaryotic protein Bag-1 may be the functional equivalent of GrpE. Bag-1 is discussed as a nucleotide release factor of Hsp70 (Bimston et al. 1998; Hohfeld and Jentsch 1997) and regulates the activity of Hsp70 proteins by binding to their amino-terminal ATPase domain (Nollen et al. 2000). Snl1 from *Saccharomyces cerevisiae* may be the yeast homologue of Bag-1 (Sondermann et al. 2002).

In addition to cochaperones, binding of short peptides also stimulates ATP hydrolysis in Hsp70 proteins by a factor of about 3. BiP, which is a member of the Hsp70 family in the ER, is one of the most extensively studied Hsp70 members in this context. Consistent with findings for DnaK (McKay 1993; Theyssen et al. 1996), it has been shown for different peptides that ATP hydrolysis is not necessary for dissociation of BiP-peptide complexes, but that the release is achieved by a conformational change upon ATP-binding (Vidal et al. 1996; Wei et al. 1995). It is not clear if the results retrieved from studies with small hydrophobic peptides can be generalized for the interaction of BiP with natural substrate proteins as well. A complex between an antibody domain and BiP showed that this substrate protein decelerates the hydrolysis step of the ATPase cycle (Mayer et al. 2003).

A low molecular weight compound called 15-deoxyspergualin (DSG) was discovered as an Hsc70-binding agent (Nadler et al. 1992). DSG binds Hsc70 with moderate affinity (K_D=4 μM) and stimulates the steady state ATPase activity of bovine and yeast cytosolic Hsp70s by 42% and 22%, respectively, but has no effect on BiP (Brodsky 1999). Possibly, DSG binds to the carboxy-terminal EEVD motif in Hsc70, as the ER-residing Hsp70s do not share this motif. Interestingly, the slight stimulation of the ATPase activity of Hsp70 is sufficient to obtain therapeutic effects. Specifically, DSG was found to reduce tissue rejection in transplant patients which may occur through the inhibition of macrophage function and induction of cytolytic T and B cells (Hughes et al. 1996).

Chaperone activity of Hsp70

The role of Hsp70s in the folding of proteins can be divided into three related activities: the promotion of folding to the native state, the prevention of aggregation, and the solubilization and refolding of aggregated proteins. In vivo, it takes about one minute to synthesize a 300-residue protein in eukaryotes (Bergmann and Lodish 1979; Waldron et al. 1974).

During this time period, many nascent chains expose hydrophobic side chains and are therefore prone to aggregation. This reaction is enhanced by a process called "macromolecular crowding" due to the highly crowded nature of the cytosol (100–400 g/liter of proteins in *E. coli*; Minton et al. 1992). Here, excluded volume effects are predicted to favor aggregation of nonnative protein chains substantially by increasing their effective concentrations (van den Berg et al. 1999). This leads to the requirement for chaperones which help to maintain nascent chains in a nonaggregated, folding-competent conformation. The mechanism by which Hsp70 chaperones assist the folding of nonnative substrates is still not elucidated in detail. Hsp70-dependent protein folding in vitro occurs typically on the time scale of minutes. Substrates undergo cycles of chaperone-bound and -free states regulated by the Hsp70 ATPase activity until the ensemble of molecules has reached the native state (Buchberger et al. 1996; Szabo et al. 1994). Chaperone binding may not only block intermolecular aggregation directly by shielding the interactive surfaces of nonnative polypeptides, including unassembled protein subunits, but may also prevent or reverse intramolecular misfolding. Hsp70s act by holding nascent and newly synthesized chains in a state competent for folding upon release into the medium. Whereas the Hsp70 proteins in bacteria and higher eukaryotes act both co- and posttranslationally, yeast and other fungi have cytosolic Hsp70 homologues that are specialized in nascent chain binding, e.g., proteins of the Ssb and Ssz class in the yeast *Saccharomyces cerevisiae*. These specialized Hsp70 proteins bind directly to the ribosome near the polypeptide exit site and are positioned to interact generally with nascent chains (Gautschi et al. 2002; Pfund et al. 1998). Whereas the majority of small proteins is able to fold rapidly and without further assistance upon completion of synthesis and release from the first set of Hsp70 chaperones, some proteins interact subsequently with a second class of nascent chain binding chaperones, including the classical Hsp70s. This class does not associate directly with the ribosome (Deuerling et al. 1999; Teter et al. 1999; Thulasiraman et al. 1999) but assists in co- or posttranslational folding or facilitate chain transfer to downstream chaperones (Deuerling et al. 1999; Siegers et al. 1999; Teter et al. 1999). The broad substrate spectrum of Hsp70s includes about 10% of total soluble protein (Deuerling et al. 2003). The importance of the Hsp70 chaperone system in vivo has been demonstrated using model substrate proteins. In this context, the enzyme ornithine transcarbamoylase has been shown to accumulate in a misfolded but soluble form in vivo when expressed in Hsp70 (Ssa)-deficient yeast (Kim et al. 1998). Firefly luciferase has been used as another model substrate for studying chaperone requirements (Frydman et al. 1992, 1994; Nimmesgern and Hartl 1993; Schroder et al. 1993; Schumacher et al. 1994). The concerted action of DnaK, DnaJ, and GrpE results in the refolding of heat-denatured luciferase in vivo and in vitro (Schroder et al. 1993). and strongly accelerates the slow, spontaneous refolding of chemically denatured firefly luciferase (~60 kDa) in vitro (Herbst et al. 1998; Szabo et al. 1994).

The question remains how cycles of Hsp70 binding and release promote protein folding. After release from Hsp70, an unfolded chain is free to fold to its native state or rebinds. One model suggests that Hsp70s actively modulate protein structure due to its secondary amide peptide bond *cis-trans* isomerase activity (APIase; Schiene-Fischer et al. 2002). APIases selectively accelerate the *cis-trans* isomerization of nonprolyl peptide bonds. Accordingly, Hsp70s may act as a regioselective catalyst for bond rotation in polypeptides allowing maintenance of nascent and newly synthesized chains in a flexible state, poised for rapid folding upon release.

Most surprisingly, the components of the Hsp70 system are missing in certain species of archaea (Leroux 2001). How aggregation of nascent and newly synthesized polypeptides is suppressed in these cells is not clear and remains to be elucidated.

The Hsp90 chaperone system

Hsp90 is one of the most abundant proteins in unstressed cells (1%–2% of total soluble cell protein; Lai et al. 1984; Welch and Feramisco 1982). It is an ubiquitous molecular chaperone found in eubacteria and all branches of eukarya, but it is apparently absent in archaea. Whereas cytoplasmic Hsp90 is essential for viability under all conditions in eukaryotes (Borkovich et al. 1989; Cutforth and Rubin 1994), the bacterial homologue HtpG is dispensable under non heat stress conditions (Bardwell and Craig 1988; Spence et al. 1990; Spence and Georgopoulos 1989). In mammalian cells, there are two genes encoding cytosolic Hsp90 homologues, with the human Hsp90α showing 85% sequence identity to Hsp90β (Hickey et al. 1989). There is also high homology to Hsp90 from lower eukaryotes and prokaryotes: yeast Hsp90 is 60% identical to human Hsp90α (Rebbe et al. 1987) and HtpG is still 34% identical to human Hsp90α (data not shown).

The first evidence for Hsp90 being a heat shock protein was elevated expression of Hsp90 in avian cells and in *Drosophila melanogaster* under nonphysiological conditions (Compton and McCarthy 1978; Farrelly and Finkelstein 1984; Oppermann et al. 1981). In addition to the cytoplasmic Hsp90, higher eukaryotes also possess distinct isoforms localized in the ER and in mitochondria. In the ER, Grp94/Gp96 (or endoplasmin; Shiu et al. 1977; Sorger and Pelham 1987) acts as a molecular chaperone, which is involved in the maturation of proteins such as nascent immunoglobulins, MHC class II, p185^{ErbB2}, collagen, HSV-1 glycoproteins, and thyroglobulin (Chavany et al. 1996; Ferreira et al. 1994; Kuznetsov et al. 1994; Little et al. 1994; Melnick et al. 1992; Navarro et al. 1991; Nicchitta 1998; Schaiff et al. 1992). Trap1 (or Hsp75) is a mitochondrially localized Hsp90 homologue (Felts et al. 2000; Song et al. 1995) with high homology to bacterial HtpG. Its function has not yet been established. Recently, a novel member of the Hsp90 family, called Hsp90N, was found (Grammatikakis et al. 2002).

Seven members of the Hsp90 family have been discovered in the plant *Arabidopsis thaliana* (Krishna and Gloor 2001). Four members are located in the cytosol, the others are predicted to be in the plastidial, mitochondrial, and endoplasmic compartments. Their tasks seem to be quite similar to their mammalian homologues, although the higher number of cytosolic Hsp90s suggests an even wider range of different target proteins.

Structural properties of Hsp90

Hsp90 is an elongated dimer with a dissociation constant in the low nM range (Maruya et al. 1999; Richter et al. 2001). In higher eukaryotes, homodimers (α-α; β-β) exist as well as heterodimers (α-β; Nemoto et al. 1996; Perdew et al. 1993). The quaternary structure is important for the ATPase activity and associated conformational changes (see the section entitled "The Hsp90 ATPase activity").

Hsp90 consists of three major domains: a highly conserved amino-terminal ATPase domain, a middle domain, and a carboxy-terminal dimerization domain (Chadli et al. 2000; Maruya et al. 1999; Prodromou et al. 1997a; Fig. 4).

In eukaryotic Hsp90s, the amino-terminal nucleotide-binding domain is connected to the remainder of the protein by a highly charged and protease-sensitive segment that is variable both in length and composition between different species and between different isoforms in the same species. This "linker" is considerably shorter in bacterial Hsp90s and in the mammalian mitochondrial homologue Trap1 (Fig. 4). Most interestingly, this seg-

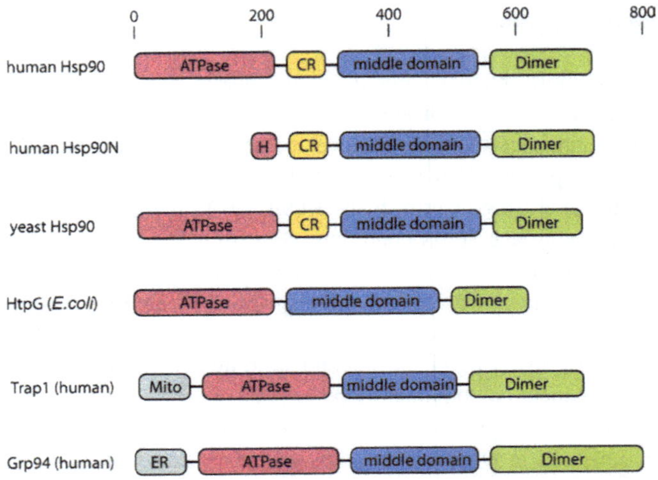

Fig. 4 Domain structure of Hsp90 family members. Comparison of the domain structures of human Hsp90, Hsp90N, yeast Hsp90, bacterial HtpG, human Trap1, and Grp94. *Pink* ATPase domain (Hsp90N: hydrophobic region), *gray* signal sequence for the endoplasmic reticulum (*ER*) or mitochondria (*MITO*), *yellow* charged region (*CR*), *blue* middle domain, *green* dimerization domain

ment is not necessary for viability in *Saccharomyces cerevisiae* (Louvion et al. 1996), which raises the question whether its function is that of a simple covalent tether, connecting the amino-terminal domain to the rest of the protein.

The three-dimensional structures of the amino-terminal domains of human and yeast Hsp90 (amino acids 1–215) have been solved by X-ray crystallography (Prodromou et al. 1997b; Stebbins et al. 1997; Fig. 5A).

The amino-terminal domain consists of a highly curved β-sheet exposed on one face, with the opposite face covered by a set of α-helices. Later, it was found that this domain contains the ATP-binding pocket (Prodromou et al. 1997a) which is identical to the binding site of the Hsp90 inhibitors geldanamycin (Stebbins et al. 1997) and radicicol (Roe et al. 1999). The structure of the amino-terminal domain of Hsp90 is similar to that of type II and type IV DNA topoisomerases, bacterial DNA gyrase B protein, and the MutL family of DNA mismatch repair proteins (Bergerat et al. 1997). The similarity to these ATP-dependent remodeling machines seems to extend to conformational changes induced by ATP binding and hydrolysis, as discussed in the section entitled "The Hsp90 ATPase activity."

The recently determined structure of the middle region of yeast Hsp90 (Meyer et al. 2003; Fig. 5B) supports the view of the evolutionary relationship between MutL and GyrB. This domain can be divided into three parts.

In the first part, residues 273–409 form a three layer α-β-α sandwich domain consisting of a five-stranded β-sheet, a three-turn α-helix, and irregular loops on the convex face. The long α-helix at the carboxy-terminus of this domain leads into a small domain (second part) formed by three short α-helices (411–420, 421–432, and 436–443) arranged in a right-handed coil. It is followed by a third domain (third part; residues 435–525) with a three-layer α-β-α sandwich architecture. The topology of the secondary structural elements in this domain is quite distinct from previously described α-β-α architectures, for example, in elongation factor G, DNA gyrase B, and the ribosomal protein S5 (Murzin 1995; Pearl and Prodromou 2000). Beyond residue 525, the structure is disordered, proba-

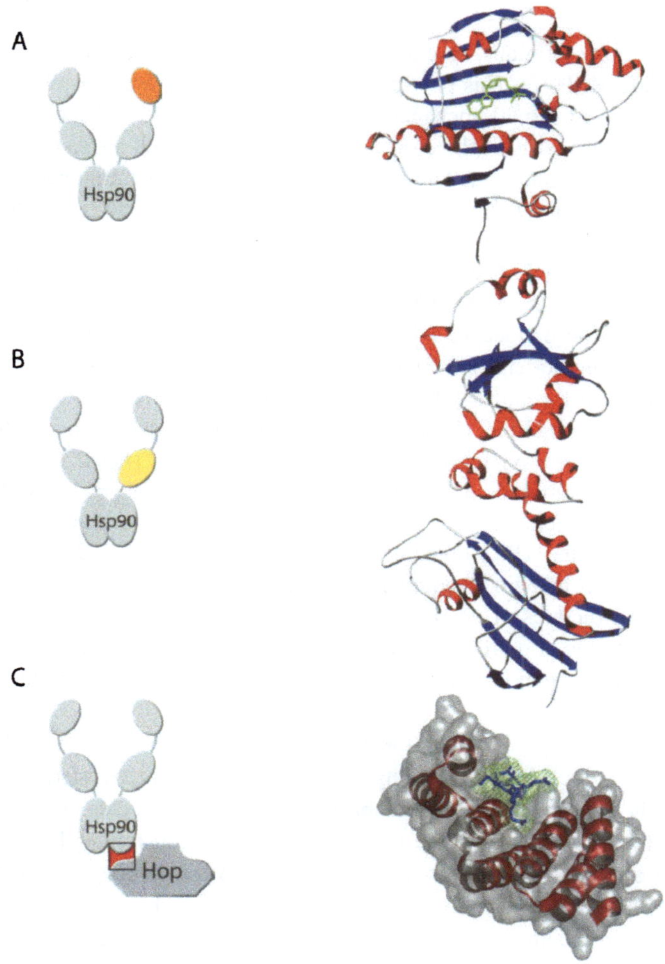

Fig. 5A–C Crystal structure of Hsp90. **A** ATPase domain of Hsp90. Crystal structure of the ATP binding site of yeast Hsp90 (PDB entry 1AM1). The bound nucleotide ADP is depicted in *green*. **B** Middle domain of Hsp90. Crystal structure of the middle domain (aa 273–560) of yeast Hsp90 (Meyer et al. 2003). **C** Binding of Hop to Hsp90. Crystal structure of the TPR2A-domain of Hop in complex with the Hsp90-peptide MEEVD (*blue*; PDB entry 1ELR)

bly reflecting flexibility in this region. In MutL and GyrB, a lysine residue in this domain serves as an acceptor of the γ-phosphate group of the bound nucleotide, and functional studies suggest that this might also be the case for Hsp90 (see the section entitled "The Hsp90 ATPase activity"). This domain may also be involved in the binding of substrate proteins.

The carboxy-terminal domain of ~12 kDa comprises a strong dimerization interface, which is essential for Hsp90 function and it provides the binding site for a subset of Hsp90 cochaperones, containing TPR domains (Chen et al. 1996b, 1998; Owens-Grillo et al. 1996b; Young et al. 1998).

Hsp90 cochaperone complexes

At the very carboxy-terminus, essentially all eukaryotic cytosolic Hsp90s terminate with the pentapeptide MEEVD. This peptide constitutes the core of the Hsp90 interaction surface for the TPRs of Hsp90 cochaperones (Ramsey et al. 2000; Scheufler et al. 2000). Each TPR-domain consists of three or more TPR motifs, which are loosely conserved repeats of roughly 34 amino acids known to mediate protein-protein interactions (Honore et al. 1992; Sikorski et al. 1990). A TPR motif shows a helix-turn-helix structure and subsequent TPR motifs are ordered in antiparallel α-helices (Das et al. 1998). In Hop, TPR domains serve to link Hsp70 and Hsp90 (Chang et al. 1997; Chen et al. 1996b; Smith and Toft 1993). Hop contains nine TPR motifs, which form two TPR domains (TPR1 and TPR2). TPR2 can be divided into TPR2A and TPR2B, each of which consists of three TPR motifs. The amino-terminal TPR-domain (TPR1) of Hop interacts with Hsp70, and the TPR-domains in the carboxy-terminal half (especially TPR2A) seem to be important for the interaction with Hsp90 (Scheufler et al. 2000). It was shown that TPR1 recognizes specifically the 12 carboxy-terminal amino acids of human Hsp70 (K_D=15 μM), and TPR2 the five carboxy-terminal amino acids of human Hsp90 (K_D=11 μM). Studies of the mouse Hop-Hsc70 interaction suggest that a network of interactions not only between charged residues in the TPR1 domain of mouse Hop and the EEVD motif of Hsc70 but also outside the TPR domain is required (Odunuga et al. 2003).

So far, only structures of single TPR domains in complex with Hsp90 are available. The crystal structures of domains of Hop in complex with short peptides from Hsp70 and Hsp90 show that the peptides are bound in an extended conformation in the TPR grooves (Fig. 5C).

Only side chains from two of the seven α-helices have contact with the peptide (Scheufler et al. 2000). Peptide binding is based mainly on electrostatic interactions between the TPR helices and the peptide backbone. The amino acids amino-terminal from this motif interact via hydrophobic and van der Waals contacts with the TPR domain (Scheufler et al. 2000). The main chain carboxylate of the carboxy-terminal aspartate of the bound peptide can be regarded as a "two-carboxylate anchor," as it is fixed with three H-bonds to the side chain groups of Lys-8, Asn-12, and Asn-43 of TPR1. The hydrophobic contacts amino-terminal of the EEVD-motif are important for the specificity of the respective bound peptide (Scheufler et al. 2000). The binding of Sti1 to the carboxy-terminal parts of Hsp70 and Hsp90 does not explain the regulatory function of Sti1 on their ATPase activities (Prodromou et al. 1999; Wegele et al. 2003a) and therefore additional contacts are postulated in the amino-terminal parts of the respective proteins.

The Hsp90 ATPase activity

ATP hydrolysis seems to be of crucial importance for Hsp90 function in vivo, because mutants that do not hydrolyze ATP do not support the functions of Hsp90 essential for viability (Obermann et al. 1998; Panaretou et al. 1998).

ATP binding to Hsp90 is generally weak with a dissociation constant in the high micromolar range (Prodromou et al. 1997a; Scheibel et al. 1997; Wegele et al. 2003b; Weikl et al. 2000). The crystal structure of the amino-terminal domain of Hsp90 in complex with ADP showed that, in contrast to most other ATP hydrolyzing proteins, ATP is bound in an unusually kinked conformation (Prodromou et al. 1997a; Fig. 5A): the adenosine ring and the ribose unit are buried in the interior of the binding pocket and the phosphate groups point to the surface. Few other proteins with weak homology to Hsp90 share this motif, like DNA Gyrase B (Wigley et al. 1991), the DNA mismatch repair protein MutL (Ban and Yang 1998), and the bacterial histidine kinase CheA (Bilwes et al. 1999). While the β-phosphate is still to some extent shielded by the binding site, the γ-phosphate of ATP seems to be completely exposed to the solution in the isolated amino-terminal domain. For GyrB, MutL, and CheA, it was shown that this γ-phosphate group is, after a structural reorganization of the domains, complexed by a lysine residue from an accompanying domain (Dutta and Inouye 2000). The kinetic analysis of the ATPase reaction showed that, also for Hsp90, regions outside the ATP-binding domain are important for efficient hydrolysis (Wegele et al. 2003b; Weikl et al. 2000). It is therefore reasonable to assume that once ATP is bound, the amino-terminal domain interacts with other parts of the Hsp90 molecule which may include an acceptor for the γ-phosphate group of ATP. The recently solved structure of the middle domain of yeast Hsp90 suggests that a possible candidate for an acceptor of the γ-phosphate group could be Lys 387 (Meyer et al. 2003).

The ATPase activity of yeast Hsp90 is weak with ~0.3 ATP hydrolyzed per minute at physiological temperature (Panaretou et al. 1998; Scheibel et al. 1998; Weikl et al. 2000). The ATPase activity of human Hsp90 is barely detectable with a k_{cat} of 0.089±0.004 min^{-1} and a K_m of 840±60 μM (McLaughlin et al. 2002; Wegele et al. 2003a; Young and Hartl 2000). Kinetic dissection of the ATPase mechanism revealed that the hydrolysis rate is determined by slow conformational changes prior to hydrolysis and a generally slow hydrolysis reaction. These changes trap the ATP molecule in Hsp90 and commit it to hydrolysis (Wegele et al. 2003b; Weikl et al. 2000). During the ATPase cycle, a weak dimerization site within the amino-terminal part of the protein becomes exposed, leading to the transient dimerization of the amino-terminal domains (Prodromou et al. 2000; Richter et al. 2002). Especially, the first 24 amino acids of this domain seem to be important for the association of the amino-terminal domains and the stimulation of ATP hydrolysis (Richter et al. 2002). Hydrolysis of ATP is tightly coupled to conformational changes of the protein, with the amino-terminal domains performing one round of association-dissociation per two molecules of ATP hydrolyzed.

The dimeric nature of the protein is of crucial importance for the hydrolyzing activity, and for the association of the amino-terminal domains during the ATPase cycle which is required for maximum ATPase activity (Prodromou et al. 2000). Deletion constructs lacking either the carboxy-terminal or both the carboxy-terminal and the middle domain with an artificially introduced Cys residue at the carboxy-terminal ends, which allows covalent dimerization, were used to analyze the contribution of the different domains and the oligomeric state on the progression of the ATPase cycle of yeast Hsp90 (Wegele et al. 2003b). All monomeric constructs tested exhibited reduced ATPase activity and a decreased affinity for ATP in comparison to wt-Hsp90. The covalently linked dimers lacking only the carboxy-terminal domain hydrolyze ATP as efficiently as the wild-type protein. Furthermore, this construct is able to trap the ATP molecule similar to the full-length protein. This demonstrates that in the ATPase cycle the carboxy-terminal domain can be replaced by a cystine bridge. In contrast, the ATPase activity of the artificially linked amino-terminal domains remains very low and bound ATP is not trapped (Wegele et al. 2003b). For chicken Hsp90α, it had been shown that a monomeric, carboxy-terminal truncation construct (aa 1–573) leads to a 100-fold increase in the ATPase activity (Owen et al. 2002). In this case, the carboxy-terminal part of the protein seems to exhibit a negative regulatory effect on the ATPase activity.

Recently, the existence of a second, low affinity nucleotide binding site in Hsp90 has been suggested, which does not differ between ATP and GTP (Garnier et al. 2002; Marcu et al. 2000; Soti et al. 2002). It is assumed to be located in the carboxy-terminal domain and to open up when the amino-terminal site is occupied.

Modulation of the Hsp90 ATPase activity

Inhibitors

Geldanamycin (GA) and the related Herbimycin A are fungal ansamycins exhibiting antitumor activity (Whitesell et al. 1992). Originally, this activity was thought to result from a direct inhibition of protein kinases such as v-Src, Raf-1, or ErbB2 (Fukazawa et al. 1991; Schnur et al. 1995a, b; Taniguchi et al. 1993). Subsequently, it turned out that GA did not inhibit these kinases directly but reduced their active cellular levels and promoted their degradation (Miller et al. 1994; Mimnaugh et al. 1996; Schneider et al. 1996; Schulte et al. 1995). Interestingly, the direct target of GA is Hsp90 and the reduction in the levels of the target kinases resulted from disruption of their productive interaction with Hsp90 (Chavany et al. 1996; Miller et al. 1994; Whitesell et al. 1994). GA binds to the amino-terminal domain of Hsp90 (Grenert et al. 1997; Roe et al. 1999; Stebbins et al. 1997) as a competitive inhibitor of ATP with a K_D of 1.2 μM (Roe et al. 1999). Thus, the affinity of GA for Hsp90 is about 500-fold higher than that of ATP (Whitesell et al. 1992, 1994).

The kinked binding conformation of ATP bound to Hsp90 is perfectly mimicked by GA (Roe et al. 1999). Based on the crystal structures of the ATPase domain in the liganded state, no differences in the conformations between the various ligands could be observed. Recent studies using NMR spectroscopy, however, suggest that specific conformational changes occur upon binding of the various ligands inside the binding pocket, but also in regions of the amino-terminal domain which are not in direct contact with the ligand (Dehner et al. 2003).

Radicicol (or monorden), which is an even more competitive inhibitor of ATP binding with nanomolar (19 nM) affinity, is a fungal macrolactone unrelated to geldanamycin (Roe et al. 1999). Crystallographic studies showed that radicicol also binds to the amino-terminal ATP binding pocket of Hsp90 (Roe et al. 1999). The drug novobiocin blocks not only ATP binding to the carboxy-terminal site, but also to the amino-terminal domain (Soti et al. 2002).

The ability of two quite distinct classes of natural compounds to bind to the ATP-binding site in Hsp90 with high specificity and affinity together with the identification of a number of key anticancer target proteins as Hsp90 clients has led to the idea of Hsp90 as an antitumor drug target. Target validation and clinical trials of geldanamycin derivatives are underway in several countries (Clarke et al. 2000; Goetz et al. 2003), e.g., the GA derivative 17-AAG (17-allylamino,17-demethoxygeldanamycin), which shows potent activity against several cancers at low nanomolar concentrations in early clinical trials (Chiosis et al. 2003)

Cochaperones, accessory proteins and modifications

As in the case of Hsp70, the ATPase cycle of Hsp90 is modulated by partner proteins which act in complex with Hsp90 in vivo. As outlined above, in higher eukaryotes, partner proteins were identified which are of fundamental importance for the Hsp90 chaperone cycle (Chen et al. 1996b; Smith et al. 1993). This mechanism is evolutionary conserved since most of the Hsp90 partner proteins involved in this cycle are known to exist in yeast as well as in man. In yeast, Sti1 seems to be a key regulator of both the Hsp70 and Hsp90 ATPases. It had been shown that Sti1 acts as a potent inhibitor of the yHsp90 ATPase (Prodromou et al. 1999). In addition to its interaction with the C-terminal end of Hsp90, Sti1 has a second binding site in the amino-terminal ATPase domain of Hsp90. Binding inhibits the amino-terminal dimerization reaction, but not nucleotide binding to Hsp90. Sti1 is a noncompetitive inhibitor which blocks the Hsp90 ATPase cycle in its nucleotide-bound state (Richter et al. 2003).

In the chaperone cycle, Sti1 is replaced by p23 and one of the large prolylisomerases. p23 (Sba1 in yeast) is a small protein with chaperone activity (Bose et al. 1996; Freeman and Morimoto 1996; Sullivan et al. 2002) that, like Sti1, interacts with the amino-terminal domain of Hsp90. In the crystal structure of human p23 (Weaver et al. 2000), the carboxy-terminal tail is unresolved, which implies that it may be unstructured. This tail is not needed for the binding of p23 to Hsp90 or to complexes with the progesterone receptor. However, the tail is necessary for optimum active chaperoning of the progesterone receptor, as well as the passive chaperoning activity of p23 in assays measuring inhibition of heat-induced protein aggregation (Weaver et al. 2000; Weikl et al. 1999).

ATP and p23 drive rearrangements in Hsp90 heterocomplexes which are important for the progression of the chaperone cycle. This interaction of p23 with Hsp90 is dependent on ATP binding (Johnson et al. 1996; Johnson and Toft 1995; Prodromou et al. 2000; Weaver et al. 2000; Young and Hartl 2000) and inhibits the inherent ATPase activity of Hsp90 (Panaretou et al. 2002). Binding is blocked by the Hsp90 inhibitors GA and radicicol, which behave as ADP mimetics (Prodromou et al. 1997b; Stebbins et al. 1997). For the p23 homologue in yeast, called Sba1, amino-terminal dimerization of Hsp90 greatly increases the affinity of Sba1 for yeast Hsp90 (Panaretou et al. 2002; and K. Richter, unpublished data). The p23 interaction is counteracted by Sti1, which prevents amino-terminal dimerization of yeast Hsp90 and therefore binding of p23 to yeast Hsp90 (K. Richter, unpublished data).

Recently, the Hch1/Aha1 proteins were identified as a new family of Hsp90 cochaperones (Panaretou et al. 2002). The prototype of this Aha1 family, Hch1, was discovered as a high copy number suppressor of an Hsp90 mutant allele, Glu381Lys (Nathan and Lindquist 1995). The location of this residue suggested that the middle segment of Hsp90 might be involved in direct interaction with Hch1 and/or Aha1. This was confirmed by coimmunoprecipitation and isothermal titration calorimetry (Meyer et al. 2003). Interestingly, Aha1 is a potent activator, accelerating the ATPase activity of yeast Hsp90 to about 12 times its basal level. The Aha1-stimulated ATPase activity of Hsp90 is sensitive to Sti1 (Panaretou et al. 2002). Unlike TPR domain cochaperones, Aha1 does not compete for binding to the carboxy-terminal domain of Hsp90 and can coexist in complexes with Sti1 and Cpr6. The TPR-domain prolylisomerase Cpr6 has no significant direct effect on the ATPase of Hsp90 (Prodromou et al. 1999), but the addition of saturating levels of Cpr6 to Hsp90-Aha1 complexes nearly doubles the Aha1-stimulated ATPase activity (Panaretou et al. 2002), suggesting a synergistic interaction of these proteins when bound to Hsp90. However, the precise function of Aha1 in the Hsp90 chaperone cycle is still unresolved.

Recently, Cdc37 (p50) has been shown to inhibit the inherent ATPase activity of Hsp90 (Siligardi et al. 2002). The suppressed ATPase activity of Hsp90 is restored when Cdc37 is displaced by the prolylisomerases Cpr6/Cyp40. Unlike Sti1, which can displace GA upon binding to Hsp90, Cdc37 forms a stable complex with geldanamycin-bound Hsp90 (Siligardi et al. 2002). However, the binding site of Cdc37 is not the same as that for TPR proteins, but it may be either very close to or overlapping with that for TPR proteins (Owens-Grillo et al. 1996a, b; Silverstein et al. 1998).

Substrate proteins

Recently, it was shown that a natural substrate influences the Hsp90 ATPase activity. Remarkably, the weak intrinsic ATPase activity of human Hsp90 could be increased up to 200-fold by the ligand-binding domain of the human glucocorticoid receptor (GR-LBD; McLaughlin et al. 2002). The stimulation seems to be highly specific, because unfolded or partially folded proteins, which are known to bind to Hsp90, did not affect the ATPase activity (McLaughlin et al. 2002). This is in contrast to most other chaperones studied so far. The stimulation of the ATPase activity by the GR-LBD affects the function of several cofactors. Hop, which has little effect on the basal ATPase activity of human Hsp90, shows a significant inhibition of the substrate-stimulated rate (McLaughlin et al. 2002). Similarly, FKBP52, a high molecular mass prolylisomerase containing a carboxy-terminal TPR domain, exhibited a stimulatory effect on the ATPase only in the presence of the GR-LBD. In the case of p23 the substrate-stimulated rate is dramatically inhibited, with a reduction in the ATPase activity to near basal levels (McLaughlin et al. 2002). However, Hsp90 homologues from *Crithidia fasciculata* and *Trypanosoma cruzi* have been shown to be stimulated up to fivefold by peptides of 6–24 amino acids (Nadeau et al. 1992).

Taken together, these results suggest that the effect of the Hsp90 cochaperones is specifically linked to substrate protein processing. Whether this is a direct consequence of conformational changes in Hsp90 upon substrate binding or whether it results from the interaction of the cochaperones with Hsp90 and the nonnative protein remains to be seen.

Hsp90 substrate proteins and chaperone function

It is usually thought that chaperone substrate proteins must be partially unfolded and that exposed hydrophobic regions are the binding sites for chaperones. However, there is no indication that steroid receptors are unfolded before their activation by the Hsp70/Hsp90 chaperone machinery. In the case of the glucocorticoid receptor, the binding site seems to be on the surface of the ligand-binding domain at the opening of the hydrophobic steroid-binding cleft (Giannoukos et al. 1999; Xu et al. 1998). Given the fact that for other substrates such as citrate synthase the interaction with highly structured folding intermediates has been demonstrated (Jakob et al. 1995), it is tempting to speculate that the Hsp90 chaperone machinery functions on proteins in their native-like conformations rather than promoting folding of denatured proteins. The client proteins of Hsp90 include steroid hormone receptors (Joab et al. 1984), helix-loop-helix transcription factors (Wilhelmsson et al. 1990), tyrosine and serine/threonine kinases (Aligue et al. 1994; Cutforth and Rubin

1994; Dai et al. 1996; Oppermann et al. 1981; Stancato et al. 1993), and the tumor suppressor p53 (Chen et al. 1996a; Sepehrnia et al. 1996), among others (Table 1). Most of these are either regulators of transcription or protein kinases. This indicates that the function of Hsp90 might be of special importance for signaling pathways and regulation. However, some of the reported examples like actin (Koyasu et al. 1986), tubulin (Sanchez et al. 1988), and proteasomes (Tsubuki et al. 1994; Wagner and Margolis 1995), do not fit this scheme. The reduction of cellular Hsp90 levels using anti-Hsp90 hammerhead ribozyme leads to accelerated cell lysis. This demonstrates the importance of Hsp90 for the maintenance of cellular integrity (Sreedhar et al. 2003).

In contrast to other chaperones, we know many substrate proteins of Hsp90 because they form stable and long-lived complexes with Hsp90 which allowed isolation, e.g., by immunoprecipitation. It has been shown that the ability of Hsp90 to prevent the aggregation of denatured proteins in vitro does not require ATP or additional protein components (Freeman and Morimoto 1996; Jakob et al. 1995; Miyata and Yahara 1992). The chaperone activity seems to map to more than one domain. At least two domains with distinct substrate specificities can prevent protein aggregation and hold substrates in a folding-competent state (Johnson et al. 2000; Minami et al. 2001; Scheibel et al. 1998; Young et al. 1997). The carboxy-terminal domain comprising the last 100 amino acids is unique in binding to partially folded proteins in an ATP-independent way, whereas the amino-terminal domain contains a peptide binding site that seems to bind preferentially peptides longer than 10 amino acids (Scheibel et al. 1998, 1999; Young et al. 1997).

For the glucocorticoid receptor (GR), it is known that Hsp90 interacts within the ligand binding domain of the receptor (Howard et al. 1990). Due to the limited stability of GR, it is suggested that the role of Hsp90 and associated chaperones is to stabilize a specific conformational state of the receptor which binds to the ligand with high affinity (McLaughlin et al. 2002).

Another class of model substrates of Hsp90 are transcription factors like the tumor suppressor protein p53. p53 contains large unstructured regions in its native state (Bell et al. 2002; Dawson et al. 2003) which makes it a candidate for interaction with molecular chaperones. For mutated p53 it has been reported that it interacts with Hsp90 in vitro and in vivo (Blagosklonny et al. 1996; Whitesell et al. 1998). Recently, it was shown that in the presence of Hop and Hsp70, Hsp90 also binds to wild-type p53 (Zylicz et al. 2001). Here again, Hsp90 does not seem to be responsible for the initial folding of p53, but rather for late-state activation and/or to keep p53 in safe custody.

Another important example for Hsp90 target proteins are the Src kinases. Neoplastic transformation of avian or mammalian cells by the Rous sarcoma virus results directly from the protein-tyrosine kinase activity of v-Src (Brugge and Erikson 1977). c-Src, the cellular counterpart from which v-Src is derived, is involved in the control of cell growth and cellular architecture (Erpel and Courtneidge 1995; Hunter 1987a, b). v-Src has a much higher kinase activity than c-Src and differs from its homologue by eight amino acid substitutions and a small carboxy-terminal replacement (Murphy et al. 1993). Newly synthesized v-Src in vertebrate cells can be coimmunoprecipitated with Hsp90 (Adkins et al. 1982; Brugge et al. 1981, 1986; Lipsich et al. 1982; Oppermann et al. 1981). For c-Src, a weaker interaction with Hsp90 was observed (Whitesell et al. 1994). However, also for the maturation of c-Src kinase, Hsp90 is necessary (Bijlmakers and Marsh 2000; Xu et al. 1999). After its attachment to the plasma membrane, Src is active and no longer associated with Hsp90 (Courtneidge and Bishop 1982; Xu and Lindquist 1993). Hsp90 binds within the kinase domain of Src (Xu et al. 1999), but Hsp90 does not appear to play a direct role in membrane association. Src kinase is only a paradigm showing the importance of Hsp90 in the maturation process of kinases. Studies with other kinases like Pkr and Hri, both members of the eIF-2α kinase family (Donze et al. 2001; Shao et al. 2001), highlight the general function of Hsp90 in these processes.

The Hsp70/Hsp90 chaperone cycle—revisited

A decade after the first description of the Hsp70/Hsp90 multichaperone cycle by David Smith (Smith 1993), it can be concluded that all basic principles of his model were confirmed with biochemical and genetic methods. The developments in biochemistry and structural biology in recent years have allowed new mechanistic questions to be addressed. Of central importance is the connection of the Hsp70 and Hsp90 components (Fig. 6). The protein Hop and its yeast homologue Sti1 are the link between the Hsp70 and Hsp90 chaperones in the mammalian and yeast system, respectively. The analysis of their mode of action revealed differences in the regulation of the mammalian Hsp90 system compared to the yeast complex (Wegele et al. 2003a). Whereas in the mammalian system Hop has no influence on the ATPase activity of the Hsp70 or Hsp90 component (Johnson et al. 1998; Wegele et al. 2003a), Sti1 is a noncompetitive inhibitor of yHsp90 (Prodromou et al. 1999; Richter et al. 2003) and a potent activator of yHsp70 (Wegele et al. 2003a). Despite sequence and functional homology of Sti1 and Hop (37% identity), there are fundamental differences between these two proteins. Concerning the ATPase activity in the absence of

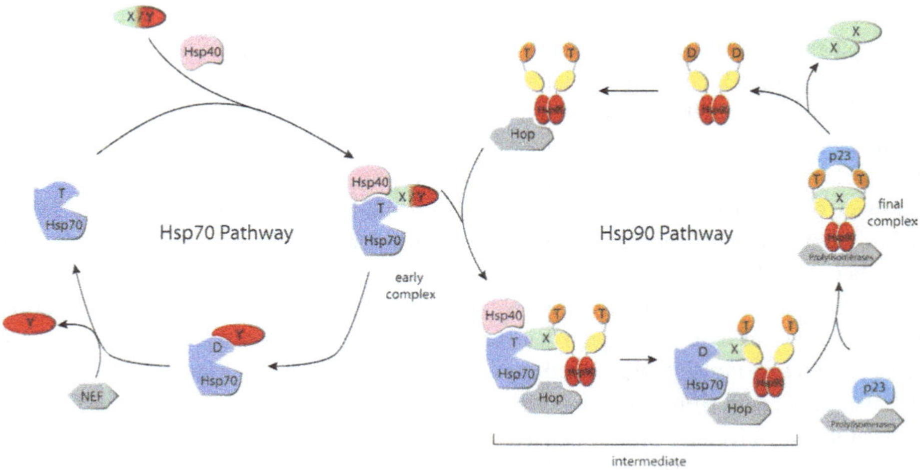

Fig. 6 Connecting the Hsp90 and Hsp70 pathways. Newly synthesized and/or partially unfolded proteins (*X/Y*) are delivered to the Hsp70 complex via Hsp40. Two possible pathways branch from this "early complex." In most cases, Hsp70 is able to process the substrate (*Y*) on its own. After ATP hydrolysis, the nucleotide exchange factor (*NEF*) releases substrate and ADP, and Hsp70 is ready to reenter the cycle. There are certain substrates (*X*) which require Hsp90 for proper folding or activation. In this case, the scaffold protein Hop connects elements of the Hsp70 and Hsp90 machineries to form the "intermediate complex." The Hsp70 component dissociates and, at the same time, p23 and prolylisomerases enter the complex. After that, the substrate is released from this "final complex." After binding to Hop, Hsp90 is able to reenter the cycle. *T* ATP-bound form, *D* ADP-bound form

substrate proteins, Hop seems to act as a passive adaptor between the Hsp70 and Hsp90 component, whereas Sti1 is a regulatory linker protein for yHsp70 and yHsp90, capable of activating the Hsp70 component and inhibiting the Hsp90 component at the same time (Wegele et al. 2003a).

This raises the question how the regulation of Hsp70 and Hsp90 is accomplished in the context of the human Hsp90 system. Interestingly, the basic ATPase activity of hHsp90 is tenfold weaker compared to yHsp90, and the ATPase activity of Hsp70 is several-fold higher compared to yHsp70. Thus, compared to the yeast system, Hsp70 seems to be already activated and Hsp90 inhibited. Therefore, a different mode of "switching" the human components "on" and "off" may be required for the human system.

Another important question is the role of the Hsp40 component and p23 in the context of the Hsp70/Hsp90 multichaperone cycle (Fig. 6). It is possible to assemble receptor-Hsp90 heterocomplexes with a minimal system of five purified proteins—Hsp90, Hsp70, Hop, Hsp40, and p23 (Dittmar et al. 1998; Kosano et al. 1998). The Hsp90-Hop-Hsp70-Hsp40 complex can be formed spontaneously from the purified proteins. The first step requires Hsp40 and ATP to form primed GR-Hsp70-Hsp40 complexes that can be activated in a second step (Dittmar and Pratt 1997). In some cases, J-domain proteins are thought to interact with substrates directly to target the subsequent binding of Hsp70 (Kelley 1998; Morishima et al. 2000b). It has been shown that Hsp40 binds directly to the progesterone receptor (PR) in the absence of Hsp70 with a stoichiometry of one molecule of Hsp40 to one molecule of receptor. Hsp40 remains bound to PR in a static manner during the assembly process (Hernandez et al. 2002a). Hsp90 and Hsp70 are sufficient for folding, but Hop, the Hsp70 partner Ydj1 and p23 accelerate, potentiate, and stabilize folding (Morishima et al. 2000a). Thus, glucocorticoid receptor function is only partially affected in yeast cells lacking Sti1 (Hop; Chang et al. 1997) or Sba1 (p23; Bohen 1998), or in cells with mutant Ydj1 (Johnson and Craig 2000). Evidence regarding the precise role of p23 remains confusing. While it is clearly associated with many mature Hsp90-substrate heterocomplexes (Forsythe et al. 2001; Hu and Seeger 1996), it may join Hsp90-substrate complexes "merely" to stimulate substrate release (Young and Hartl 2000), bind substrates independently or even after the release from Hsp90 (Freeman et al. 2000; Hu et al. 1997). Recent studies showed that p23 localizes in vivo to genomic response elements in a hormone-dependent manner, disrupting receptor-mediated transcriptional activation in vivo and in vitro (Freeman and Yamamoto 2002). This suggests that molecular chaperones promote disassembly of transcriptional regulatory complexes, thus enabling regulatory machineries to detect and respond to signaling changes.

Another basic principle addressed in the last years is the interaction between the components of the multichaperone cycle via TPR domains. Most cochaperones contain and utilize TPRs to bind Hsp90. Concerning the Hsp70/Hsp90 chaperone network, this is the case for Hop (Sti1 in yeast), the prolylisomerases FKBP51, FKBP52, cyclophilin-40 (Cpr6 and Cpr7 in yeast), and protein phosphatase 5 (Ppt1 in yeast). These interactions depend on the carboxy-terminal MEEVD of Hsp90, but additional sequences elsewhere in Hsp90 modulate these interactions (Carrello et al. 1999; Chen et al. 1998; Prodromou et al. 1999; Ramsey et al. 2000; Richter et al. 2003). Interestingly, this pentapeptide is dispensable for viability of *Saccharomyces cerevisiae* (Louvion et al. 1996). With the exception of Cns1, also the TPR-containing Hsp90 cochaperones are not essential for viability in yeast (Duina et al. 1996; Marsh et al. 1998; Nicolet and Craig 1989; Warth et al. 1997) (Table 2).

Together with Hop, Hsp70 is also released from the complex. Dissociation of Hop allows binding of TPR-domain-prolylisomerases such as FKBP52, FKBP51, PP5, or CyP-40. The role of the prolylisomerase components in the cycle in the case of SHRs is beginning to be understood. The binding of hormone seems to induce the substitution of one (FKBP51) for another (FKBP52) and the transport protein dynein is recruited. FKBP52 selectively potentiates hormone-dependent reporter gene activation by as much as 20-fold at limiting hormone concentrations (Riggs et al. 2003). Furthermore, binding of hormone to the receptor induces the translocation of the hormone-generated GR-Hsp90-FKBP52-dynein complex from the cytoplasm to the nucleus, a step that precedes dissociation of the complex within the nucleus and conversion of GR to the DNA-binding form (Davies et al. 2002).

Besides the precise role of the classical components of the multichaperone cycle, the function of additional cofactors, e.g., Cdc37, needs to be elucidated. Cdc37 is a cochaperone that is required for Hsp90 substrate-specific folding activity. Cdc37 is essential in yeast (Gerber et al. 1995; Lee et al. 2002) and genetically linked to several protein kinases (MacLean and Picard 2003). It is required for the production of Cdc28/cyclin complexes (Gerber et al. 1995) and appears to target Hsp90 to the Cdk4 kinase (Dai et al. 1996; Lamphere et al. 1997; Stepanova et al. 1996). Cdc37 binds immature protein kinases through interaction with its amino-terminal region (Grammatikakis et al. 1999), and links these to Hsp90 via its carboxy-terminal region (Shao et al. 2001). In addition to a physical role in targeting kinases to Hsp90 for activation, yeast Cdc37 also exhibits chaperone activity (Kimura et al. 1997). Cdc37 is critical not only for kinase activation but also for their stability. Blocking Cdc37 function in immortalized cell lines using either a dominant negative Cdc37 mutant (Grammatikakis et al. 1999) or inhibiting Hsp90 activity with GA (Whitesell et al. 1994) results in decreases in steady-state protein levels and in the activity of client kinases, such as Cdk4, Raf-1, v-Src, Akt, and the androgen receptor.

Hsp90 and Cdc37 have been coimmunoprecipitated with the kinases v-Src, c-Src, Fes, and Raf (Brugge et al. 1981, 1986; Nair et al. 1996; Oppermann et al. 1981; Stancato et al. 1993; Wartmann and Davis 1994). Cdc37 also binds to Hsp90 in the absence of kinases (Whitelaw et al. 1991), which implies that these two proteins may form a core chaperone machinery for folding of proteins like v-Src and c-Src, which belong to the nonreceptor tyrosine kinases.

In the recently identified novel member of the Hsp90 family, called Hsp90N (Grammatikakis et al. 2002), the ansamycin-binding domain is replaced by a much shorter, hydrophobic sequence, preceded by a putative myristylation signal. Although much less abundant, Hsp90N binds Raf with a higher affinity than Hsp90. In sharp contrast to Hsp90, Hsp90N does not associate with Cdc37.

Perspectives

In recent years, a steady progress in the analysis of the structural and functional mechanisms of Hsp90 and its cochaperones has been revealed. New functions and new cochaperones have been emerging—"stagnation" is not in sight. The properties of some new cofactors and regulators of the Hsp90 and Hsp70 chaperones such as Aha1 or Cdc37 raises the question if there is only "one" Hsp90 multichaperone cycle. It is more likely that several multichaperone cycles exist, which work independently and in parallel and which differ in cofactor and substrate composition. We assume that substrate proteins select specific Hsp90 cofactors to guarantee proper folding and activation.

In this context, it was shown that Hsp90 must be phosphorylated, or rephosphorylated in vitro by casein kinase II, to stimulate the proper maturation of the heme-regulated inhibitor Hri (Szyszka et al. 1989). For the Hsp90 substrates v-Src and the retrovirus protein σ1, phosphorylation of Hsp90 seems to trigger substrate release (Mimnaugh et al. 1995; Zhao et al. 2001). However, genetic evidence supports the idea that the protein kinase B equivalent Sch9 acts as a repressor of Hsp90 function in yeast (Morano and Thiele 1999). There is mounting evidence that Hsp90 function is influenced by phosphorylation. However, the critical phosphorylation sites, the kinases that act in intact cells and the molecular mechanisms are unknown. Thus, the role of Hsp90 phosphorylation still needs to be elucidated.

The involvement of Hsp90 in many signaling and regulatory pathways suggests a capacitory role for morphological evolution by buffering cryptic genetic variation (Queitsch et al. 2002; Rutherford and Lindquist 1998). In the fruit fly *Drosophila melanogaster* and the plant *Arabidopsis thalianana*, it was shown that mutation or pharmacological inhibition of Hsp90 can produce morphological changes affecting the structure of the respective organism. Thus, Hsp90 is believed to maintain the “normal” behavior of mutated signaling proteins whose altered phenotype becomes manifest only when Hsp90 function is comprised. This role in “buffering” seems to be conserved across the plant and animal kingdoms.

Despite a growing list of Hsp90 substrates, substrate recognition and specificity remain poorly understood. As in vivo studies with budding yeast have revealed that most cellular proteins do not need Hsp90 for folding under physiological conditions (Nathan et al. 1997), it remains to be elucidated which factors distinguish Hsp90-dependent from Hsp90-independent proteins. One hint could be that most of the proteins, which require Hsp90, are labile and undergo rapid turnover. In these cases, Hsp90 seems to act more as a “holdase” or as an “activator” than as a folding machine. For substrates, which are prone to degradation, another TPR protein, called CHIP, is important. It is able to bind Hsc70 and Hsp90 and accelerates ubiquitin-dependent degradation of chaperone substrates (Ballinger et al. 1999; Connell et al. 2001; Jiang et al. 2001). Thus, Hsp90 may also be a key player in determining the final fate of its targets.

Regarding the Hsp90 mechanism, functional domains and ATP-dependent domain movements should be defined more clearly. The structure of the entire molecule as well as its conformational changes driven by ATP and the binding of cochaperones and substrates need to be elucidated. For most Hsp90 cochaperones, including p23 and Cdc37, the interaction surfaces have not been determined. Even for TPR proteins, we do not know exactly how the interaction surface extends beyond the carboxy-terminal MEEVD sequence.

Progress has been made on the development of new biochemical, genetic, and pharmacological tools. The invention of highly specific Hsp90 inhibitors has revolutionized both basic and applied research on Hsp90. The use of Hsp90 drugs to treat certain types of cancer (Ochel et al. 2001; Piper 2001) will certainly benefit from a better understanding of Hsp90 functions at the level of whole cells and organisms. A recent study shows that Hsp90 derived from tumor cells has a 100-fold higher binding affinity for the geldanamycin derivative 17-AAG than Hsp90 from normal cells (Kamal et al. 2003). This result suggests that tumor cells contain Hsp90 complexes in an activated, high-affinity conformation that facilitates malignant progression, which may represent a unique target for cancer therapeutics.

This unexpected finding highlights the present state of the field. We have witnessed amazing progress in many aspects of the Hsp90 chaperone machinery. However, some basic questions remain to be elucidated, especially the understanding of the Hsp90 chaperone machine at the molecular and organismic level.

Acknowledgements. Financial support by the DFG to JB is gratefully acknowledged. Harald Wegele was supported by fellowships of the Studienstiftung des Deutschen Volkes and the Fonds der Chemischen Industrie. We thank Otmar Hainzl, Jochen Reinstein and Thomas Scheibel for comments on the manuscript.

References

Abbas-Terki T, Picard D (1999) Alpha-complemented beta-galactosidase. An in vivo model substrate for the molecular chaperone heat-shock protein 90 in yeast. Eur J Biochem 266:517–523

Abravaya K, Myers MP, Murphy SP, Morimoto RI (1992) The human heat shock protein hsp70 interacts with HSF, the transcription factor that regulates heat shock gene expression. Genes Dev 6:1153–1164

Adkins B, Hunter T, Sefton BM (1982) The transforming proteins of PRCII virus and Rous sarcoma virus form a complex with the same two cellular phosphoproteins. J Virol 43:448–455

Agarraberes FA, Dice JF (2001) A molecular chaperone complex at the lysosomal membrane is required for protein translocation. J Cell Sci 114:2491–2499

Ali A, Bharadwaj S, O'Carroll R, Ovsenek N (1998) HSP90 interacts with and regulates the activity of heat shock factor 1 in Xenopus oocytes. Mol Cell Biol 18:4949–4960

Aligue R, Akhavan-Niak H, Russell P (1994) A role for Hsp90 in cell cycle control: Weel tyrosine kinase activity requires interaction with Hsp90. EMBO J 13:6099–6106

An WG, Schulte TW, Neckers LM (2000) The heat shock protein 90 antagonist geldanamycin alters chaperone association with p210bcr-abl and v-src proteins before their degradation by the proteasome. Cell Growth Differ 11:355–360

Arbeitman MN, Hogness DS (2000) Molecular chaperones activate the Drosophila ecdysone receptor, an RXR heterodimer. Cell 101:67–77

Ballinger CA, Connell P, Wu Y, Hu Z, Thompson LJ, Yin LY, Patterson C (1999) Identification of CHIP, a novel tetratricopeptide repeat-containing protein that interacts with heat shock proteins and negatively regulates chaperone functions. Mol Cell Biol 19:4535–4545

Ban C, Yang W (1998) Crystal structure and ATPase activity of MutL: implications for DNA repair and mutagenesis. Cell 95:541–552

Banecki B, Zylicz M (1996) Real time kinetics of the DnaK/DnaJ/GrpE molecular chaperone machine action. J Biol Chem 271:6137–6143

Banumathy G, Singh V, Tatu U (2002) Host chaperones are recruited in membrane-bound complexes by Plasmodium falciparum. J Biol Chem 277:3902–3912

Bardwell JC, Craig EA (1988) Ancient heat shock gene is dispensable. J Bacteriol 170:2977–2983

Beckmann RP, Mizzen LE, Welch WJ (1990) Interaction of Hsp 70 with newly synthesized proteins: implications for protein folding and assembly. Science 248:850–854

Bell S, Klein C, Muller L, Hansen S, Buchner J (2002) p53 contains large unstructured regions in its native state. J Mol Biol 322:917–927

Ben Or S (1989) Evidence that 5 S intermediate state in glucocorticoid receptor transformation contains hsp90 in addition to the steroid-binding protein. J Steroid Biochem 33:899–906

Benaroudj N, Batelier G, Triniolles F, Ladjimi MM (1995) Self-association of the molecular chaperone HSC70. Biochemistry 34:15282–15290

Bender AT, Silverstein AM, Demady DR, Kanelakis KC, Noguchi S, Pratt WB, Osawa Y (1999) Neuronal nitric-oxide synthase is regulated by the Hsp90-based chaperone system in vivo. J Biol Chem 274:1472–1478

Bergerat A, de Massy B, Gadelle D, Varoutas PC, Nicolas A, Forterre P (1997) An atypical topoisomerase II from Archaea with implications for meiotic recombination. Nature 386:414–417

Bergmann JE, Lodish HF (1979) A kinetic model of protein synthesis. Application to hemoglobin synthesis and translational control. J Biol Chem 254:11927–11937

Bernstein SL, Russell P, Wong P, Fishelevich R, Smith LE (2001) Heat shock protein 90 in retinal ganglion cells: association with axonally transported proteins. Vis Neurosci 18:429–436

Bijlmakers MJ, Marsh M (2000) Hsp90 is essential for the synthesis and subsequent membrane association, but not the maintenance, of the Src-kinase p56(lck). Mol Biol Cell 11:1585–1595

Bilwes AM, Alex LA, Crane BR, Simon MI (1999) Structure of CheA, a signal-transducing histidine kinase. Cell 96:131–141

Bimston D, Song J, Winchester D, Takayama S, Reed JC, Morimoto RI (1998) BAG-1, a negative regulator of Hsp70 chaperone activity, uncouples nucleotide hydrolysis from substrate release. EMBO J 17:6871–6878

Blagosklonny MV, Toretsky J, Bohen S, Neckers L (1996) Mutant conformation of p53 translated in vitro or in vivo requires functional HSP90. Proc Natl Acad Sci 93:8379–8383

Blond-Elguindi S, Cwirla SE, Dower WJ, Lipshutz RJ, Sprang SR, Sambrook JF, Gething MJ (1993a) Affinity panning of a library of peptides displayed on bacteriophages reveals the binding specificity of BiP. Cell 75:717–728

Blond-Elguindi S, Fourie AM, Sambrook JF, Gething MJ (1993b) Peptide-dependent stimulation of the ATPase activity of the molecular chaperone BiP is the result of conversion of oligomers to active monomers. J Biol Chem 268:12730–12735

Bogatcheva NV, Ma Y, Urosev D, Gusev NB (1999) Localization of calponin binding sites in the structure of 90-kDa heat shock protein (Hsp90). FEBS Lett 457:369–374

Bohen SP (1998) Genetic and biochemical analysis of p23 and ansamycin antibiotics in the function of Hsp90-dependent signaling proteins. Mol Cell Biol 18:3330–3339

Bolliger L, Deloche O, Glick BS, Georgopoulos C, Jeno P, Kronidou N, Horst M, Morishima N, Schatz G (1994) A mitochondrial homolog of bacterial GrpE interacts with mitochondrial hsp70 and is essential for viability. EMBO J 13:1998–2006

Borkovich KA, Farrelly FW, Finkelstein DB, Taulien J, Lindquist S (1989) hsp82 is an essential protein that is required in higher concentrations for growth of cells at higher temperatures. Mol Cell Biol 9:3919–3930

Bose S, Weikl T, Bugl H, Buchner J (1996) Chaperone function of Hsp90-associated proteins. Science 274:1715–1717

Brehmer D, Rudiger S, Gassler CS, Klostermeier D, Packschies L, Reinstein J, Mayer MP, Bukau B (2001) Tuning of chaperone activity of Hsp70 proteins by modulation of nucleotide exchange. Nat Struct Biol 8:427–432

Brodsky JL (1999) Selectivity of the molecular chaperone-specific immunosuppressive agent 15-deoxyspergualin: modulation of Hsc70 ATPase activity without compromising DnaJ chaperone interactions. Biochem Pharmacol 57:877–880

Brugge JS (1986) Interaction of the Rous sarcoma virus protein pp60src with the cellular proteins pp50 and pp90. Curr Top Microbiol Immunol 123:1–22

Brugge JS, Erikson RL (1977) Identification of a transformation-specific antigen induced by an avian sarcoma virus. Nature 269:346–348

Brugge JS, Erikson E, Erikson RL (1981) The specific interaction of the Rous sarcoma virus transforming protein, pp60src, with two cellular proteins. Cell 25:363–372

Bruneau N, Lombardo D, Bendayan M (1998) Participation of GRP94-related protein in secretion of pancreatic bile salt-dependent lipase and in its internalization by the intestinal epithelium. J Cell Sci 111 (Pt 17):2665–2679

Buchberger A, Schroder H, Hesterkamp T, Schonfeld HJ, Bukau B (1996) Substrate shuttling between the DnaK and GroEL systems indicates a chaperone network promoting protein folding. J Mol Biol 261:328–333

Buchner J (1999) Hsp90 & Co.–a holding for folding. Trends Biochem Sci 24:136–141

Bukau B, Horwich AL (1998) The Hsp70 and Hsp60 chaperone machines. Cell 92:351–366

Busconi L, Guan J, Denker BM (2000) Degradation of heterotrimeric Galpha(o) subunits via the proteosome pathway is induced by the hsp90-specific compound geldanamycin. J Biol Chem 275:1565–1569

Carrello A, Ingley E, Minchin RF, Tsai S, Ratajczak T (1999) The common tetratricopeptide repeat acceptor site for steroid receptor-associated immunophilins and hop is located in the dimerization domain of Hsp90. J Biol Chem 274:2682–2689

Catelli MG, Binart N, Jung-Testas I, Renoir JM, Baulieu EE, Feramisco JR, Welch WJ (1985) The common 90-kd protein component of nontransformed "8S" steroid receptors is a heat-shock protein. EMBO J 4:3131–3135

Chadli A, Bouhouche I, Sullivan W, Stensgard B, McMahon N, Catelli MG, Toft DO (2000) Dimerization and N-terminal domain proximity underlie the function of the molecular chaperone heat shock protein 90. Proc Natl Acad Sci 97:12524–12529

Chang GC, Liu R, Panniers R, Li GC (1994) Rat fibroblasts transfected with the human 70-kDa heat shock gene exhibit altered translation and eukaryotic initiation factor 2 alpha phosphorylation following heat shock. Int J Hyperthermia 10:325–337

Chang HC, Lindquist S (1994) Conservation of Hsp90 macromolecular complexes in Saccharomyces cerevisiae. J Biol Chem 269:24983–24988

Chang HC, Nathan DF, Lindquist S (1997) In vivo analysis of the Hsp90 cochaperone Sti1 (p60). Mol Cell Biol 17:318–325

Chavany C, Mimnaugh E, Miller P, Bitton R, Nguyen P, Trepel J, Whitesell L, Schnur R, Moyer J, Neckers L (1996) p185erbB2 binds to GRP94 in vivo. Dissociation of the p185erbB2/GRP94 heterocomplex by benzoquinone ansamycins precedes depletion of p185erbB2. J Biol Chem 271:4974–4977

Cheetham ME, Caplan AJ (1998) Structure, function and evolution of DnaJ: conservation and adaptation of chaperone function. Cell Stress Chaperones 3:28–36

Chen CF, Chen Y, Dai K, Chen PL, Riley DJ, Lee WH (1996a) A new member of the hsp90 family of molecular chaperones interacts with the retinoblastoma protein during mitosis and after heat shock. Mol Cell Biol 16:4691–4699

Chen S, Prapapanich V, Rimerman RA, Honore B, Smith DF (1996b) Interactions of p60, a mediator of progesterone receptor assembly, with heat shock proteins hsp90 and hsp70. Mol Endocrinol 10:682–693

Chen S, Sullivan WP, Toft DO, Smith DF (1998) Differential interactions of p23 and the TPR-containing proteins Hop, Cyp40, FKBP52 and FKBP51 with Hsp90 mutants. Cell Stress Chaperones 3:118–129

Chiosis G, Huezo H, Rosen N, Mimnaugh E, Whitesell L, Neckers L (2003) 17AAG: low target binding affinity and potent cell activity–finding an explanation. Mol Cancer Ther 2:123–129

Clarke PA, Hostein I, Banerji U, Stefano FD, Maloney A, Walton M, Judson I, Workman P (2000) Gene expression profiling of human colon cancer cells following inhibition of signal transduction by 17-allylamino-17-demethoxygeldanamycin, an inhibitor of the hsp90 molecular chaperone. Oncogene 19:4125–4133

Compton JL, McCarthy BJ (1978) Induction of the Drosophila heat shock response in isolated polytene nuclei. Cell 14:191–201

Connell P, Ballinger CA, Jiang J, Wu Y, Thompson LJ, Hohfeld J, Patterson C (2001) The cochaperone CHIP regulates protein triage decisions mediated by heat-shock proteins. Nat Cell Biol 3:93–96

Courtneidge SA, Bishop JM (1982) Transit of pp60v-src to the plasma membrane. Proc Natl Acad Sci 79:7117–7121

Crevel G, Bates H, Huikeshoven H, Cotterill S (2001) The Drosophila Dpit47 protein is a nuclear Hsp90 cochaperone that interacts with DNA polymerase alpha. J Cell Sci 114:2015–2025

Cutforth T, Rubin GM (1994) Mutations in Hsp83 and cdc37 impair signaling by the sevenless receptor tyrosine kinase in Drosophila. Cell 77:1027–1036

Cyr DM, Douglas MG (1994) Differential regulation of Hsp70 subfamilies by the eukaryotic DnaJ homologue YDJ1. J Biol Chem 269:9798–9804

Cyr DM, Lu X, Douglas MG (1992) Regulation of Hsp70 function by a eukaryotic DnaJ homolog. J Biol Chem 267:20927–20931

Dai K, Kobayashi R, Beach D (1996) Physical interaction of mammalian CDC37 with CDK4. J Biol Chem 271:22030–22034

Dalman FC, Bresnick EH, Patel PD, Perdew GH, Watson SJ, Jr., Pratt WB (1989) Direct evidence that the glucocorticoid receptor binds to hsp90 at or near the termination of receptor translation in vitro. J Biol Chem 264:19815-19821

Das AK, Cohen PW, Barford D (1998) The structure of the tetratricopeptide repeats of protein phosphatase 5: implications for TPR-mediated protein-protein interactions. EMBO J 17:1192–1199

Davies TH, Ning YM, Sanchez ER (2002) A new first step in activation of steroid receptors: hormone-induced switching of FKBP51 and FKBP52 immunophilins. J Biol Chem 277:4597–4600

Dawson R, Muller L, Dehner A, Klein C, Kessler H, Buchner J (2003) The N-terminal domain of p53 is natively unfolded. J Mol Biol 332:1131–1141

de Carcer G, do Carmo AM, Lallena MJ, Glover DM, Gonzalez C (2001) Requirement of Hsp90 for centrosomal function reflects its regulation of Polo kinase stability. EMBO J 20:2878–2884

Dehner A, Furrer J, Richter K, Schuster I, Buchner J, Kessler H (2003) NMR chemical shift perturbation study of the N-terminal domain of Hsp90 upon binding of ADP, AMP-PNP, geldanamycin, and radicicol. Chembiochem 4:870–877

Dekker PJ, Pfanner N (1997) Role of mitochondrial GrpE and phosphate in the ATPase cycle of matrix Hsp70. J Mol Biol 270:321–327

Denis M, Cuthill S, Wikstrom AC, Poellinger L, Gustafsson JA (1988) Association of the dioxin receptor with the Mr 90,000 heat shock protein: a structural kinship with the glucocorticoid receptor. Biochem Biophys Res Commun 155:801–807

Denis M, Gustafsson JA (1989) Translation of glucocorticoid receptor mRNA in vitro yields a nonactivated protein. J Biol Chem 264:6005–6008

Deshaies RJ, Koch BD, Werner-Washburne M, Craig EA, Schekman R (1988) A subfamily of stress proteins facilitates translocation of secretory and mitochondrial precursor polypeptides. Nature 332:800–805

Deuerling E, Patzelt H, Vorderwulbecke S, Rauch T, Kramer G, Schaffitzel E, Mogk A, Schulze-Specking A, Langen H, Bukau B (2003) Trigger factor and DnaK possess overlapping substrate pools and binding specificities. Mol Microbiol 47:1317–1328

Deuerling E, Schulze-Specking A, Tomoyasu T, Mogk A, Bukau B (1999) Trigger factor and DnaK cooperate in folding of newly synthesized proteins. Nature 400:693–696

Dittmar KD, Banach M, Galigniana MD, Pratt WB (1998) The role of DnaJ-like proteins in glucocorticoid receptor hsp90 heterocomplex assembly by the reconstituted hsp90.p60.hsp70 foldosome complex. J Biol Chem 273:7358–7366

Dittmar KD, Pratt WB (1997) Folding of the glucocorticoid receptor by the reconstituted Hsp90-based chaperone machinery. The initial hsp90.p60.hsp70-dependent step is sufficient for creating the steroid binding conformation. J Biol Chem 272:13047–13054

Donze O, Abbas-Terki T, Picard D (2001) The Hsp90 chaperone complex is both a facilitator and a repressor of the dsRNA-dependent kinase PKR. EMBO J 20:3771–3780

Donze O, Picard D (1999) Hsp90 binds and regulates Gcn2, the ligand-inducible kinase of the alpha subunit of eukaryotic translation initiation factor 2 [corrected]. Mol Cell Biol 19:8422–8432

Duina AA, Marsh JA, Gaber RF (1996) Identification of two CyP-40-like cyclophilins in Saccharomyces cerevisiae, one of which is required for normal growth. Yeast 12:943–952

Dutta R, Inouye M (2000) GHKL, an emergent ATPase/kinase superfamily. Trends Biochem Sci 25:24–28

Erpel T, Courtneidge SA (1995) Src family protein tyrosine kinases and cellular signal transduction pathways. Curr Opin Cell Biol 7:176–182

Evans RM, Hollenberg SM (1988) Cooperative and positional independent trans-activation domains of the human glucocorticoid receptor. Cold Spring Harb Symp Quant Biol 53 Pt 2:813–818

Farrelly FW, Finkelstein DB (1984) Complete sequence of the heat shock-inducible HSP90 gene of Saccharomyces cerevisiae. J Biol Chem 259:5745–5751

Felts SJ, Owen BA, Nguyen P, Trepel J, Donner DB, Toft DO (2000) The hsp90-related protein TRAP1 is a mitochondrial protein with distinct functional properties. J Biol Chem 275:3305–3312

Ferreira LR, Norris K, Smith T, Hebert C, Sauk JJ (1994) Association of Hsp47, Grp78, and Grp94 with procollagen supports the successive or coupled action of molecular chaperones. J Cell Biochem 56:518–526

Fink AL (1997) Molecular chaperones in the life cycle of proteins. Marcel Dekker, New York, pp 123–150

Fisher DL, Mandart E, Doree M (2000) Hsp90 is required for c-Mos activation and biphasic MAP kinase activation in Xenopus oocytes. EMBO J 19:1516–1524

Flaherty KM, DeLuca-Flaherty C, McKay DB (1990) Three-dimensional structure of the ATPase fragment of a 70 K heat-shock cognate protein. Nature 346:623–628

Flaherty KM, McKay DB, Kabsch W, Holmes KC (1991) Similarity of the three-dimensional structures of actin and the ATPase fragment of a 70-kDa heat shock cognate protein. Proc Natl Acad Sci 88:5041–5045

Flanagan CA, Thorner J (1992) Purification and characterization of a soluble phosphatidylinositol 4-kinase from the yeast Saccharomyces cerevisiae. J Biol Chem 267:24117–24125

Fletterick RJ, Bates DJ, Steitz TA (1975) The structure of a yeast hexokinase monomer and its complexes with substrates at 2.7-A resolution. Proc Natl Acad Sci 72:38–42

Flynn GC, Pohl J, Flocco MT, Rothman JE (1991) Peptide-binding specificity of the molecular chaperone BiP. Nature 353:726–730

Forsythe HL, Jarvis JL, Turner JW, Elmore LW, Holt SE (2001) Stable association of hsp90 and p23, but Not hsp70, with active human telomerase. J Biol Chem 276:15571–15574

Freeman BC, Felts SJ, Toft DO, Yamamoto KR (2000) The p23 molecular chaperones act at a late step in intracellular receptor action to differentially affect ligand efficacies. Genes Dev 14:422–434

Freeman BC, Morimoto RI (1996) The human cytosolic molecular chaperones hsp90, hsp70 (hsc70) and hdj-1 have distinct roles in recognition of a nonnative protein and protein refolding. EMBO J 15:2969–2979

Freeman BC, Toft DO, Morimoto RI (1996) Molecular chaperone machines: chaperone activities of the cyclophilin Cyp-40 and the steroid aporeceptor-associated protein p23. Science 274:1718–1720

Freeman BC, Yamamoto KR (2002) Disassembly of transcriptional regulatory complexes by molecular chaperones. Science 296:2232–2235
Frydman J, Nimmesgern E, Erdjument-Bromage H, Wall JS, Tempst P, Hartl FU (1992) Function in protein folding of TRiC, a cytosolic ring complex containing TCP-1 and structurally related subunits. EMBO J 11:4767–4778
Frydman J, Nimmesgern E, Ohtsuka K, Hartl FU (1994) Folding of nascent polypeptide chains in a high molecular mass assembly with molecular chaperones. Nature 370:111–117
Fujita N, Sato S, Ishida A, Tsuruo T (2002) Involvement of Hsp90 in signaling and stability of 3-phosphoinositide-dependent kinase-1. J Biol Chem 277:10346–10353
Fukazawa H, Li PM, Yamamoto C, Murakami Y, Mizuno S, Uehara Y (1991) Specific inhibition of cytoplasmic protein tyrosine kinases by herbimycin A in vitro. Biochem Pharmacol 42:1661–1671
Gamer J, Bujard H, Bukau B (1992) Physical interaction between heat shock proteins DnaK, DnaJ, and GrpE and the bacterial heat shock transcription factor sigma 32. Cell 69:833–842
Gao B, Emoto Y, Greene L, Eisenberg E (1993) Nucleotide binding properties of bovine brain uncoating ATPase. J Biol Chem 268:8507–8513
Gao B, Greene L, Eisenberg E (1994) Characterization of nucleotide-free uncoating ATPase and its binding to ATP, ADP, and ATP analogues. Biochemistry 33:2048–2054
Garcia-Cardena G, Fan R, Shah V, Sorrentino R, Cirino G, Papapetropoulos A, Sessa WC (1998) Dynamic activation of endothelial nitric oxide synthase by Hsp90. Nature 392:821–824
Garnier C, Lafitte D, Tsvetkov PO, Barbier P, Leclerc-Devin J, Millot JM, Briand C, Makarov AA, Catelli MG, Peyrot V (2002) Binding of ATP to heat shock protein 90: evidence for an ATP-binding site in the C-terminal domain. J Biol Chem 277:12208–12214
Gautschi M, Lilie H, Funfschilling U, Mun A, Ross S, Lithgow T, Rucknagel P, Rospert S (2001) RAC, a stable ribosome-associated complex in yeast formed by the DnaK-DnaJ homologs Ssz1p and zuotin. Proc Natl Acad Sci 98:3762–3767
Gautschi M, Mun A, Ross S, Rospert S (2002) A functional chaperone triad on the yeast ribosome. Proc Natl Acad Sci 99:4209–4214
Gerber MR, Farrell A, Deshaies RJ, Herskowitz I, Morgan DO (1995) Cdc37 is required for association of the protein kinase Cdc28 with G1 and mitotic cyclins. Proc Natl Acad Sci 92:4651–4655
Giannoukos G, Silverstein AM, Pratt WB, Simons SS, Jr. (1999) The seven amino acids (547–553) of rat glucocorticoid receptor required for steroid and hsp90 binding contain a functionally independent LXXLL motif that is critical for steroid binding. J Biol Chem 274:36527–36536
Gilmore R, Coffey MC, Lee PW (1998) Active participation of Hsp90 in the biogenesis of the trimeric reovirus cell attachment protein sigma1. J Biol Chem 273:15227–15233
Goes FS, Martin J (2001) Hsp90 chaperone complexes are required for the activity and stability of yeast protein kinases Mik1, Wee1 and Swe1. Eur J Biochem 268:2281–2289
Goetz MP, Toft DO, Ames MM, Erlichman C (2003) The Hsp90 chaperone complex as a novel target for cancer therapy. Ann Oncol 14:1169–1176
Gragerov A, Zeng L, Zhao X, Burkholder W, Gottesman ME (1994) Specificity of DnaK-peptide binding. J Mol Biol 235:848–854
Grammatikakis N, Lin JH, Grammatikakis A, Tsichlis PN, Cochran BH (1999) p50(cdc37) acting in concert with Hsp90 is required for Raf-1 function. Mol Cell Biol 19:1661–1672
Grammatikakis N, Vultur A, Ramana CV, Siganou A, Schweinfest CW, Watson DK, Raptis L (2002) The role of Hsp90N, a new member of the Hsp90 family, in signal transduction and neoplastic transformation. J Biol Chem 277:8312–8320
Greene LE, Zinner R, Naficy S, Eisenberg E (1995) Effect of nucleotide on the binding of peptides to 70-kDa heat shock protein. J Biol Chem 270:2967–2973
Grenert JP, Sullivan WP, Fadden P, Haystead TA, Clark J, Mimnaugh E, Krutzsch H, Ochel HJ, Schulte TW, Sausville E, Neckers LM, Toft DO (1997) The amino-terminal domain of heat shock protein 90 (hsp90) that binds geldanamycin is an ATP/ADP switch domain that regulates hsp90 conformation. J Biol Chem 272:23843–23850
Gusarova V, Caplan AJ, Brodsky JL, Fisher EA (2001) Apoprotein B degradation is promoted by the molecular chaperones hsp90 and hsp70. J Biol Chem 276:24891–24900
Ha JH, Hellman U, Johnson ER, Li L, McKay DB, Sousa MC, Takeda S, Wernstedt C, Wilbanks SM (1997) Destabilization of peptide binding and interdomain communication by an E543 K mutation in the bovine 70-kDa heat shock cognate protein, a molecular chaperone. J Biol Chem 272:27796–27803
Ha JH, Johnson ER, McKay DB, Sousa MC, Takeda S, Wilbanks SM (1999) Structure and mechanism of Hsp70 proteins. In: Molecular chaperones and folding catalysts:regulation, cellular function and mechanisms. Harwood Academic Publishers, pp 573–608

Ha JH, McKay DB (1994) ATPase kinetics of recombinant bovine 70-kDa heat shock cognate protein and its amino-terminal ATPase domain. Biochemistry 33:14625–14635

Ha JH, McKay DB (1995) Kinetics of nucleotide-induced changes in the tryptophan fluorescence of the molecular chaperone Hsc70 and its subfragments suggest the ATP-induced conformational change follows initial ATP binding. Biochemistry 34:11635–11644

Hansen WJ, Lingappa VR, Welch WJ (1994) Complex environment of nascent polypeptide chains. J Biol Chem 269:26610–26613

Harrison CJ, Hayer-Hartl M, Di Liberto M, Hartl F, Kuriyan J (1997) Crystal structure of the nucleotide exchange factor GrpE bound to the ATPase domain of the molecular chaperone DnaK. Science 276:431–435

Hartson SD, Matts RL (1994) Association of Hsp90 with cellular Src-family kinases in a cell-free system correlates with altered kinase structure and function. Biochemistry 33:8912–8920

Hashimoto Y, Shudo K (1991) Cytosolic-nuclear tumor promoter-specific binding protein: association with the 90-kDa heat shock protein and translocation into nuclei by treatment with 12-O-tetradecanoylphorbol 13-acetate. Jpn J Cancer Res 82:665–675

Hendrick JP, Hartl FU (1993) Molecular chaperone functions of heat-shock proteins. Annu Rev Biochem 62:349–384

Herbertsson H, Kuhme T, Hammarstrom S (1999) The 650-kDa 12(S)-hydroxyeicosatetraenoic acid binding complex: occurrence in human platelets, identification of hsp90 as a constituent, and binding properties of its 50-kDa subunit. Arch Biochem Biophys 367:33–38

Herbst R, Gast K, Seckler R (1998) Folding of firefly (Photinus pyralis) luciferase: aggregation and reactivation of unfolding intermediates. Biochemistry 37:6586–6597

Hernandez MP, Chadli A, Toft DO (2002a) HSP40 binding is the first step in the HSP90 chaperoning pathway for the progesterone receptor. J Biol Chem 277:11873–11881

Hernandez MP, Sullivan WP, Toft DO (2002b) The assembly and intermolecular properties of the hsp70-Hop-hsp90 molecular chaperone complex. J Biol Chem 277:38294–38304

Herrmann JM, Neupert W (2000) Protein transport into mitochondria. Curr Opin Microbiol 3:210–214

Hickey E, Brandon SE, Smale G, Lloyd D, Weber LA (1989) Sequence and regulation of a gene encoding a human 89-kilodalton heat shock protein. Mol Cell Biol 9:2615–2626

Hoff KG, Silberg JJ, Vickery LE (2000) Interaction of the iron-sulfur cluster assembly protein IscU with the Hsc66/Hsc20 molecular chaperone system of Escherichia coli. Proc Natl Acad Sci 97:7790–7795

Hohfeld J, Jentsch S (1997) GrpE-like regulation of the hsc70 chaperone by the anti-apoptotic protein BAG-1. EMBO J 16:6209–6216

Holley SJ, Yamamoto KR (1995) A role for Hsp90 in retinoid receptor signal transduction. Mol Biol Cell 6:1833–1842

Holt SE, Aisner DL, Baur J, Tesmer VM, Dy M, Ouellette M, Trager JB, Morin GB, Toft DO, Shay JW, Wright WE, White MA (1999) Functional requirement of p23 and Hsp90 in telomerase complexes. Genes Dev 13:817–826

Honore B, Leffers H, Madsen P, Rasmussen HH, Vandekerckhove J, Celis JE (1992) Molecular cloning and expression of a transformation-sensitive human protein containing the TPR motif and sharing identity to the stress-inducible yeast protein STI1. J Biol Chem 267:8485–8491

Hoshino T, Wang J, Devetten MP, Iwata N, Kajigaya S, Wise RJ, Liu JM, Youssoufian H (1998) Molecular chaperone GRP94 binds to the Fanconi anemia group C protein and regulates its intracellular expression. Blood 91:4379–4386

Howard KJ, Holley SJ, Yamamoto KR, Distelhorst CW (1990) Mapping the HSP90 binding region of the glucocorticoid receptor. J Biol Chem 265:11928–11935

Hu J, Seeger C (1996) Hsp90 is required for the activity of a hepatitis B virus reverse transcriptase. Proc Natl Acad Sci 93:1060–1064

Hu J, Toft DO, Seeger C (1997) Hepadnavirus assembly and reverse transcription require a multi-component chaperone complex which is incorporated into nucleocapsids. EMBO J 16:59–68

Hughes SE, Xiao S, Perera S, Fluno C, Hasselbarth J, Jayasankar V, Singh J, Del Rosario A, Freed BM, Singh TP, Lempert N, Conti DJ, Gruber SA (1996) Local immunosuppression of canine renal allografts with 15-deoxyspergualin. Transplant Proc 28:2054–2055

Hundley H, Eisenman H, Walter W, Evans T, Hotokezaka Y, Wiedmann M, Craig E (2002) The in vivo function of the ribosome-associated Hsp70, Ssz1, does not require its putative peptide-binding domain. Proc Natl Acad Sci 99:4203–4208

Hunter T (1987a) A tail of two src's: mutatis mutandis. Cell 49:1–4

Hunter T (1987b) A thousand and one protein kinases. Cell 50:823–829

Hupp TR, Meek DW, Midgley CA, Lane DP (1992) Regulation of the specific DNA binding function of p53. Cell 71:875–886

Hurley JH, Faber HR, Worthylake D, Meadow ND, Roseman S, Pettigrew DW, Remington SJ (1993) Structure of the regulatory complex of Escherichia coli IIIGlc with glycerol kinase. Science 259:673–677

Hutchison KA, Brott BK, De Leon JH, Perdew GH, Jove R, Pratt WB (1992) Reconstitution of the multiprotein complex of pp60src, hsp90, and p50 in a cell-free system. J Biol Chem 267:2902–2908

Hutchison KA, Stancato LF, Owens-Grillo JK, Johnson JL, Krishna P, Toft DO, Pratt WB (1995) The 23-kDa acidic protein in reticulocyte lysate is the weakly bound component of the hsp foldosome that is required for assembly of the glucocorticoid receptor into a functional heterocomplex with hsp90. J Biol Chem 270:18841–18847

Imai J, Yahara I (2000) Role of HSP90 in salt stress tolerance via stabilization and regulation of calcineurin. Mol Cell Biol 20:9262–9270

Inanobe A, Takahashi K, Katada T (1994) Association of the beta gamma subunits of trimeric GTP-binding proteins with 90-kDa heat shock protein, hsp90. J Biochem.(Tokyo) 115:486–492

Itoh T, Matsuda H, Mori H (1999) Phylogenetic analysis of the third hsp70 homolog in Escherichia coli; a novel member of the Hsc66 subfamily and its possible cochaperone. DNA Res 6:299–305

Jaiswal RK, Weissinger E, Kolch W, Landreth GE (1996) Nerve growth factor-mediated activation of the mitogen-activated protein (MAP) kinase cascade involves a signaling complex containing B-Raf and HSP90. J Biol Chem 271:23626–23629

Jakob U, Lilie H, Meyer I, Buchner J (1995) Transient interaction of Hsp90 with early unfolding intermediates of citrate synthase. Implications for heat shock in vivo. J Biol Chem 270:7288–7294

Jerome V, Leger J, Devin J, Baulieu EE, Catelli MG (1991) Growth factors acting via tyrosine kinase receptors induce HSP90 alpha gene expression. Growth Factors 4:317–327

Jiang J, Ballinger CA, Wu Y, Dai Q, Cyr DM, Hohfeld J, Patterson C (2001) CHIP is a U-box-dependent E3 ubiquitin ligase: identification of Hsc70 as a target for ubiquitylation. J Biol Chem 276:42938–42944

Joab I, Radanyi C, Renoir M, Buchou T, Catelli MG, Binart N, Mester J, Baulieu EE (1984) Common non-hormone binding component in nontransformed chick oviduct receptors of four steroid hormones. Nature 308:850–853

Johnson BD, Chadli A, Felts SJ, Bouhouche I, Catelli MG, Toft DO (2000) Hsp90 chaperone activity requires the full-length protein and interaction among its multiple domains. J Biol Chem 275:32499–32507

Johnson BD, Schumacher RJ, Ross ED, Toft DO (1998) Hop modulates Hsp70/Hsp90 interactions in protein folding. J Biol Chem 273:3679–3686

Johnson J, Corbisier R, Stensgard B, Toft D (1996) The involvement of p23, hsp90, and immunophilins in the assembly of progesterone receptor complexes. J Steroid Biochem Mol Biol 56:31–37

Johnson JL, Beito TG, Krco CJ, Toft DO (1994) Characterization of a novel 23-kilodalton protein of unactive progesterone receptor complexes. Mol Cell Biol 14:1956–1963

Johnson JL, Craig EA (2000) A role for the Hsp40 Ydj1 in repression of basal steroid receptor activity in yeast. Mol Cell Biol 20:3027–3036

Johnson JL, Toft DO (1994) A novel chaperone complex for steroid receptors involving heat shock proteins, immunophilins, and p23. J Biol Chem 269:24989–24993

Johnson JL, Toft DO (1995) Binding of p23 and hsp90 during assembly with the progesterone receptor. Mol Endocrinol 9:670–678

Jolly C, Morimoto RI (2000) Role of the heat shock response and molecular chaperones in oncogenesis and cell death. J Natl Cancer Inst 92:1564–1572

Joly GA, Ayres M, Kilbourn RG (1997) Potent inhibition of inducible nitric oxide synthase by geldanamycin, a tyrosine kinase inhibitor, in endothelial, smooth muscle cells, and in rat aorta. FEBS Lett 403:40–44

Kabsch W, Mannherz HG, Suck D, Pai EF, Holmes KC (1990) Atomic structure of the actin: DNase I complex. Nature 347:37–44

Kamal A, Thao L, Sensintaffar J, Zhang L, Boehm MF, Fritz LC, Burrows FJ (2003) A high-affinity conformation of Hsp90 confers tumor selectivity on Hsp90 inhibitors. Nature 425:407–410

Kang J, Kim T, Ko YG, Rho SB, Park SG, Kim MJ, Kwon HJ, Kim S (2000) Heat shock protein 90 mediates protein-protein interactions between human aminoacyl-tRNA synthetases. J Biol Chem 275:31682–31688

Kellermayer MS, Csermely P (1995) ATP induces dissociation of the 90-kDa heat shock protein (hsp90) from F-actin: interference with the binding of heavy meromyosin. Biochem Biophys Res Commun 211:166–174

Kelley WL (1998) The J-domain family and the recruitment of chaperone power. Trends Biochem Sci 23:222–227

Kim S, Schilke B, Craig EA, Horwich AL (1998) Folding in vivo of a newly translated yeast cytosolic enzyme is mediated by the SSA class of cytosolic yeast Hsp70 proteins. Proc Natl Acad Sci 95:12860–12865

Kimura Y, Rutherford SL, Miyata Y, Yahara I, Freeman BC, Yue L, Morimoto RI, Lindquist S (1997) Cdc37 is a molecular chaperone with specific functions in signal transduction. Genes Dev 11:1775–1785

Klostermeier D, Seidel R, Reinstein J (1998) Functional properties of the molecular chaperone DnaK from Thermus thermophilus. J Mol Biol 279:841–853

Knarr G, Gething MJ, Modrow S, Buchner J (1995) BiP binding sequences in antibodies. J Biol Chem 270:27589–27594

Knarr G, Kies U, Bell S, Mayer M, Buchner J (2002) Interaction of the chaperone BiP with an antibody domain: implications for the chaperone cycle. J Mol Biol 318:611–620

Knarr G, Modrow S, Todd A, Gething MJ, Buchner J (1999) BiP-binding sequences in HIV gp160. Implications for the binding specificity of bip. J Biol Chem 274:29850–29857

Komori A, Sueoka E, Fujiki H, Ishii M, Kozu T (1999) Association of MTG8 (ETO/CDR), a leukemia-related protein, with serine/threonine protein kinases and heat shock protein HSP90 in human hematopoietic cell lines. Jpn J Cancer Res 90:60–68

Kosano H, Stensgard B, Charlesworth MC, McMahon N, Toft D (1998) The assembly of progesterone receptor-hsp90 complexes using purified proteins. J Biol Chem 273:32973–32979

Koyasu S, Nishida E, Kadowaki T, Matsuzaki F, Iida K, Harada F, Kasuga M, Sakai H, Yahara I (1986) Two mammalian heat shock proteins, HSP90 and HSP100, are actin-binding proteins. Proc Natl Acad Sci 83:8054–8058

Krishna P, Gloor G (2001) The Hsp90 family of proteins in Arabidopsis thaliana. Cell Stress Chaperones 6:238–246

Kumar R, Grammatikakis N, Chinkers M (2001) Regulation of the atrial natriuretic peptide receptor by heat shock protein 90 complexes. J Biol Chem 276:11371–11375

Kuznetsov G, Chen LB, Nigam SK (1994) Several endoplasmic reticulum stress proteins, including ERp72, interact with thyroglobulin during its maturation. J Biol Chem 269:22990–22995

Lai BT, Chin NW, Stanek AE, Keh W, Lanks KW (1984) Quantitation and intracellular localization of the 85 K heat shock protein by using monoclonal and polyclonal antibodies. Mol Cell Biol 4:2802–2810

Laloraya S, Gambill BD, Craig EA (1994) A role for a eukaryotic GrpE-related protein, Mge1p, in protein translocation. Proc Natl Acad Sci 91:6481–6485

Lamphere L, Fiore F, Xu X, Brizuela L, Keezer S, Sardet C, Draetta GF, Gyuris J (1997) Interaction between Cdc37 and Cdk4 in human cells. Oncogene 14:1999–2004

Landry SJ, Jordan R, McMacken R, Gierasch LM (1992) Different conformations for the same polypeptide bound to chaperones DnaK and GroEL. Nature 355:455–457

Lee P, Rao J, Fliss A, Yang E, Garrett S, Caplan AJ (2002) The Cdc37 protein kinase-binding domain is sufficient for protein kinase activity and cell viability. J Cell Biol 159:1051–1059

Leroux MR (2001) Protein folding and molecular chaperones in archaea. Adv Appl Microbiol 50:219–277

Lewis J, Devin A, Miller A, Lin Y, Rodriguez Y, Neckers L, Liu ZG (2000) Disruption of hsp90 function results in degradation of the death domain kinase, receptor-interacting protein (RIP), and blockage of tumor necrosis factor-induced nuclear factor-kappaB activation. J Biol Chem 275:10519–10526

Liberek K, Marszalek J, Ang D, Georgopoulos C, Zylicz M (1991) Escherichia coli DnaJ and GrpE heat shock proteins jointly stimulate ATPase activity of DnaK. Proc Natl Acad Sci 88:2874–2878

Lipsich LA, Cutt JR, Brugge JS (1982) Association of the transforming proteins of Rous, Fujinami, and Y73 avian sarcoma viruses with the same two cellular proteins. Mol Cell Biol 2:875–880

Little E, Ramakrishnan M, Roy B, Gazit G, Lee AS (1994) The glucose-regulated proteins (GRP78 and GRP94): functions, gene regulation, and applications. Crit Rev Eukaryot Gene Expr 4:1–18

Loo MA, Jensen TJ, Cui L, Hou Y, Chang XB, Riordan JR (1998) Perturbation of Hsp90 interaction with nascent CFTR prevents its maturation and accelerates its degradation by the proteasome. EMBO J 17:6879–6887

Lopez-Buesa P, Pfund C, Craig EA (1998) The biochemical properties of the ATPase activity of a 70-kDa heat shock protein (Hsp70) are governed by the C-terminal domains. Proc Natl Acad Sci 95:15253–15258

Louvion JF, Abbas-Terki T, Picard D (1998) Hsp90 is required for pheromone signaling in yeast. Mol Biol Cell 9:3071–3083

Louvion JF, Warth R, Picard D (1996) Two eukaryote-specific regions of Hsp82 are dispensable for its viability and signal transduction functions in yeast. Proc Natl Acad Sci 93:13937–13942

Lovric J, Bischof O, Moelling K (1994) Cell cycle-dependent association of Gag-Mil and hsp90. FEBS Lett 343:15–21

Lu Z, Cyr DM (1998) Protein folding activity of Hsp70 is modified differentially by the hsp40 cochaperones Sis1 and Ydj1. J Biol Chem 273:27824–27830
Lutz T, Westermann B, Neupert W, Herrmann JM (2001) The mitochondrial proteins Ssq1 and Jac1 are required for the assembly of iron sulfur clusters in mitochondria. J Mol Biol 307:815–825
Macario AJ, Lange M, Ahring BK, De Macario EC (1999) Stress genes and proteins in the archaea. Microbiol Mol Biol Rev 63:923–67
MacLean M, Picard D (2003) Cdc37 goes beyond Hsp90 and kinases. Cell Stress Chaperones 8:114–119
Mahony D, Parry DA, Lees E (1998) Active cdk6 complexes are predominantly nuclear and represent only a minority of the cdk6 in T cells. Oncogene 16:603–611
Marcu MG, Chadli A, Bouhouche I, Catelli M, Neckers LM (2000) The heat shock protein 90 antagonist novobiocin interacts with a previously unrecognized ATP-binding domain in the carboxyl terminus of the chaperone. J Biol Chem 275:37181–37186
Marsh JA, Kalton HM, Gaber RF (1998) Cns1 is an essential protein associated with the hsp90 chaperone complex in Saccharomyces cerevisiae that can restore cyclophilin 40-dependent functions in cpr7Delta cells. Mol Cell Biol 18:7353–7359
Maruya M, Sameshima M, Nemoto T, Yahara I (1999) Monomer arrangement in HSP90 dimer as determined by decoration with N and C-terminal region specific antibodies. J Mol Biol 285:903–907
Masson-Gadais B, Houle F, Laferriere J, Huot J (2003) Integrin alphavbeta3, requirement for VEGFR2-mediated activation of SAPK2/p38 and for Hsp90-dependent phosphorylation of focal adhesion kinase in endothelial cells activated by VEGF. Cell Stress Chaperones 8:37–52
Matts RL, Hurst R (1989) Evidence for the association of the heme-regulated eIF-2 alpha kinase with the 90-kDa heat shock protein in rabbit reticulocyte lysate in situ. J Biol Chem 264:15542–15547
Mayer M, Reinstein J, Buchner J (2003) Modulation of the ATPase cycle of BiP by peptides and proteins. J Mol Biol 330:137–144
Mayer MP, Brehmer D, Gassler CS, Bukau B (2001) Hsp70 chaperone machines. Adv Protein Chem 59:1–44
Mayer MP, Schroder H, Rudiger S, Paal K, Laufen T, Bukau B (2000) Multistep mechanism of substrate binding determines chaperone activity of Hsp70. Nat Struct Biol 7:586–593
McCarty JS, Buchberger A, Reinstein J, Bukau B (1995) The role of ATP in the functional cycle of the DnaK chaperone system. J Mol Biol 249:126–137
McGuire J, Coumailleau P, Whitelaw ML, Gustafsson JA, Poellinger L (1995) The basic helix-loop-helix/PAS factor Sim is associated with hsp90. Implications for regulation by interaction with partner factors. J Biol Chem 270:31353–31357
McKay DB (1993) Structure and mechanism of 70-kDa heat-shock-related proteins. Adv Protein Chem 44:67–98
McLaughlin SH, Smith HW, Jackson SE (2002) Stimulation of the weak ATPase activity of human hsp90 by a client protein. J Mol Biol 315:787–798
Melnick J, Aviel S, Argon Y (1992) The endoplasmic reticulum stress protein GRP94, in addition to BiP, associates with unassembled immunoglobulin chains. J Biol Chem 267:21303–21306
Meyer P, Prodromou C, Hu B, Vaughan C, Roe SM, Panaretou B, Piper PW, Pearl LH (2003) Structural and functional analysis of the middle segment of hsp90: implications for ATP hydrolysis and client protein and cochaperone interactions. Mol Cell 11:647–658
Milani V, Noessner E, Ghose S, Kuppner M, Ahrens B, Scharner A, Gastpar R, Issels RD (2002) Heat shock protein 70: role in antigen presentation and immune stimulation. Int J Hyperthermia 18:563–575
Miller P, Schnur RC, Barbacci E, Moyer MP, Moyer JD (1994) Binding of benzoquinoid ansamycins to p100 correlates with their ability to deplete the erbB2 gene product p185. Biochem Biophys Res Commun 201:1313–1319
Mimnaugh EG, Chavany C, Neckers L (1996) Polyubiquitination and proteasomal degradation of the p185c-erbB-2 receptor protein-tyrosine kinase induced by geldanamycin. J Biol Chem 271:22796–22801
Mimnaugh EG, Worland PJ, Whitesell L, Neckers LM (1995) Possible role for serine/threonine phosphorylation in the regulation of the heteroprotein complex between the hsp90 stress protein and the pp60v-src tyrosine kinase. J Biol Chem 270:28654–28659
Minami M, Nakamura M, Emori Y, Minami Y (2001) Both the N- and C-terminal chaperone sites of Hsp90 participate in protein refolding. Eur J Biochem 268:2520–2524
Minet E, Mottet D, Michel G, Roland I, Raes M, Remacle J, Michiels C (1999) Hypoxia-induced activation of HIF-1: role of HIF-1alpha-Hsp90 interaction. FEBS Lett 460:251–256
Minton AP, Colclasure GC, Parker JC (1992) Model for the role of macromolecular crowding in regulation of cellular volume. Proc Natl Acad Sci 89:10504–10506

Miyata Y, Ikawa Y, Shibuya M, Nishida E (2001) Specific association of a set of molecular chaperones including HSP90 and Cdc37 with MOK, a member of the mitogen-activated protein kinase superfamily. J Biol Chem 276:21841–21848
Miyata Y, Yahara I (1992) The 90-kDa heat shock protein, HSP90, binds and protects casein kinase II from self-aggregation and enhances its kinase activity. J Biol Chem 267:7042–7047
Miyata Y, Yahara I (2000) p53-independent association between SV40 large T antigen and the major cytosolic heat shock protein, HSP90. Oncogene 19:1477–1484
Mizuno K, Shirogane T, Shinohara A, Iwamatsu A, Hibi M, Hirano T (2001) Regulation of Pim-1 by Hsp90. Biochem Biophys Res Commun 281:663–669
Mogk A, Tomoyasu T, Goloubinoff P, Rudiger S, Roder D, Langen H, Bukau B (1999) Identification of thermolabile Escherichia coli proteins: prevention and reversion of aggregation by DnaK and ClpB. EMBO J 18:6934–6949
Morano KA, Thiele DJ (1999) The Sch9 protein kinase regulates Hsp90 chaperone complex signal transduction activity in vivo. EMBO J 18:5953–5962
Morishima Y, Kanelakis KC, Silverstein AM, Dittmar KD, Estrada L, Pratt WB (2000a) The Hsp organizer protein hop enhances the rate of but is not essential for glucocorticoid receptor folding by the multiprotein Hsp90-based chaperone system. J Biol Chem 275:6894–6900
Morishima Y, Murphy PJ, Li DP, Sanchez ER, Pratt WB (2000b) Stepwise assembly of a glucocorticoid receptor hsp90 heterocomplex resolves two sequential ATP-dependent events involving first hsp70 and then hsp90 in opening of the steroid binding pocket. J Biol Chem 275:18054–18060
Multhoff G (2002) Activation of natural killer cells by heat shock protein 70. Int J Hyperthermia 18:576–585
Munoz MJ, Jimenez J (1999) Genetic interactions between Hsp90 and the Cdc2 mitotic machinery in the fission yeast Schizosaccharomyces pombe. Mol Gen Genet 261:242–250
Muresan Z, Arvan P (1997) Thyroglobulin transport along the secretory pathway. Investigation of the role of molecular chaperone, GRP94, in protein export from the endoplasmic reticulum. J Biol Chem 272:26095–26102
Murphy SM, Bergman M, Morgan DO (1993) Suppression of c-Src activity by C-terminal Src kinase involves the c-Src SH2 and SH3 domains: analysis with Saccharomyces cerevisiae. Mol Cell Biol 13:5290–5300
Murzin AG (1995) A ribosomal protein module in EF-G and DNA gyrase. Nat Struct Biol 2:25–26
Nadeau K, Sullivan MA, Bradley M, Engman DM, Walsh CT (1992) 83-kilodalton heat shock proteins of trypanosomes are potent peptide-stimulated ATPases. Protein Sci 1:970–979
Nadler SG, Tepper MA, Schacter B, Mazzucco CE (1992) Interaction of the immunosuppressant deoxyspergualin with a member of the Hsp70 family of heat shock proteins. Science 258:484–486
Nair SC, Toran EJ, Rimerman RA, Hjermstad S, Smithgall TE, Smith DF (1996) A pathway of multi-chaperone interactions common to diverse regulatory proteins: estrogen receptor, Fes tyrosine kinase, heat shock transcription factor Hsf1, and the aryl hydrocarbon receptor. Cell Stress Chaperones 1:237–250
Nakai M, Kato Y, Ikeda E, Toh-e A, Endo T (1994) Yge1p, a eukaryotic Grp-E homolog, is localized in the mitochondrial matrix and interacts with mitochondrial Hsp70. Biochem Biophys Res Commun 200:435–442
Nakamura T, Hinagata J, Tanaka T, Imanishi T, Wada Y, Kodama T, Doi T (2002) HSP90, HSP70, and GAPDH directly interact with the cytoplasmic domain of macrophage scavenger receptors. Biochem Biophys Res Commun 290:858–864
Nathan DF, Lindquist S (1995) Mutational analysis of Hsp90 function: interactions with a steroid receptor and a protein kinase. Mol Cell Biol 15:3917–3925
Nathan DF, Vos MH, Lindquist S (1997) In vivo functions of the Saccharomyces cerevisiae Hsp90 chaperone. Proc Natl Acad Sci 94:12949–12956
Navarro D, Qadri I, Pereira L (1991) A mutation in the ectodomain of herpes simplex virus 1 glycoprotein B causes defective processing and retention in the endoplasmic reticulum. Virology 184:253–264
Nelson RJ, Ziegelhoffer T, Nicolet C, Werner-Washburne M, Craig EA (1992) The translation machinery and 70 kd heat shock protein cooperate in protein synthesis. Cell 71:97–105
Nemoto T, Matsusaka T, Ota M, Takagi T, Collinge DB, Walther-Larsen H (1996) Dimerization characteristics of the 94-kDa glucose-regulated protein. J Biochem.(Tokyo) 120:249–256
Nicchitta CV (1998) Biochemical, cell biological and immunological issues surrounding the endoplasmic reticulum chaperone GRP94/gp96. Curr Opin Immunol 10:103–109
Nicolet CM, Craig EA (1989) Isolation and characterization of STI1, a stress-inducible gene from Saccharomyces cerevisiae. Mol Cell Biol 9:3638–3646
Nimmesgern E, Hartl FU (1993) ATP-dependent protein refolding activity in reticulocyte lysate. Evidence for the participation of different chaperone components. FEBS Lett 331:25–30

Nollen EA, Brunsting JF, Song J, Kampinga HH, Morimoto RI (2000) Bag1 functions in vivo as a negative regulator of Hsp70 chaperone activity. Mol Cell Biol 20:1083–1088

Obermann WM, Sondermann H, Russo AA, Pavletich NP, Hartl FU (1998) In vivo function of Hsp90 is dependent on ATP binding and ATP hydrolysis. J Cell Biol 143:901–910

O'Brien MC, Flaherty KM, McKay DB (1996) Lysine 71 of the chaperone protein Hsc70 Is essential for ATP hydrolysis. J Biol Chem 271:15874–15878

O'Brien MC, McKay DB (1993) Threonine 204 of the chaperone protein Hsc70 influences the structure of the active site, but is not essential for ATP hydrolysis. J Biol Chem 268:24323–24329

O'Brien MC, McKay DB (1995) How potassium affects the activity of the molecular chaperone Hsc70. I. Potassium is required for optimal ATPase activity. J Biol Chem 270:2247–2250

Ochel HJ, Eichhorn K, Gademann G (2001) Geldanamycin: the prototype of a class of antitumor drugs targeting the heat shock protein 90 family of molecular chaperones. Cell Stress Chaperones 6:105–112

Odunuga OO, Hornby JA, Bies C, Zimmermann R, Pugh DJ, Blatch GL (2003) Tetratricopeptide repeat motif-mediated Hsc70-mSTI1 interaction. Molecular characterization of the critical contacts for successful binding and specificity. J Biol Chem 278:6896–6904

O'Keeffe B, Fong Y, Chen D, Zhou S, Zhou Q (2000) Requirement for a kinase-specific chaperone pathway in the production of a Cdk9/cyclin T1 heterodimer responsible for P-TEFb-mediated tat stimulation of HIV-1 transcription. J Biol Chem 275:279–287

Oppermann H, Levinson W, Bishop JM (1981) A cellular protein that associates with the transforming protein of Rous sarcoma virus is also a heat-shock protein. Proc Natl Acad Sci 78:1067–1071

Osipiuk J, Walsh MA, Freeman BC, Morimoto RI, Joachimiak A (1999) Structure of a new crystal form of human Hsp70 ATPase domain. Acta Crystallogr D Biol Crystallogr 55:1105–1107

Owen BA, Sullivan WP, Felts SJ, Toft DO (2002) Regulation of heat shock protein 90 ATPase activity by sequences in the carboxyl terminus. J Biol Chem 277:7086–7091

Owens-Grillo JK, Czar MJ, Hutchison KA, Hoffmann K, Perdew GH, Pratt WB (1996a) A model of protein targeting mediated by immunophilins and other proteins that bind to hsp90 via tetratricopeptide repeat domains. J Biol Chem 271:13468–13475

Owens-Grillo JK, Stancato LF, Hoffmann K, Pratt WB, Krishna P (1996b) Binding of immunophilins to the 90-kDa heat shock protein (hsp90) via a tetratricopeptide repeat domain is a conserved protein interaction in plants. Biochemistry 35:15249–15255

Packschies L, Theyssen H, Buchberger A, Bukau B, Goody RS, Reinstein J (1997) GrpE accelerates nucleotide exchange of the molecular chaperone DnaK with an associative displacement mechanism. Biochemistry 36:3417–3422

Pai KS, Mahajan VB, Lau A, Cunningham DD (2001) Thrombin receptor signaling to cytoskeleton requires Hsp90. J Biol Chem 276:32642–32647

Palleros DR, Reid KL, Shi L, Welch WJ, Fink AL (1993) ATP-induced protein-Hsp70 complex dissociation requires K+ but not ATP hydrolysis. Nature 365:664–666

Palmquist K, Riis B, Nilsson A, Nygard O (1994) Interaction of the calcium and calmodulin regulated eEF-2 kinase with heat shock protein 90. FEBS Lett 349:239–242

Panaretou B, Prodromou C, Roe SM, O'Brien R, Ladbury JE, Piper PW, Pearl LH (1998) ATP binding and hydrolysis are essential to the function of the Hsp90 molecular chaperone in vivo. EMBO J 17:4829–4836

Panaretou B, Siligardi G, Meyer P, Maloney A, Sullivan JK, Singh S, Millson SH, Clarke PA, Naaby-Hansen S, Stein R, Cramer R, Mollapour M, Workman P, Piper PW, Pearl LH, Prodromou C (2002) Activation of the ATPase activity of hsp90 by the stress-regulated cochaperone aha1. Mol Cell 10:1307–1318

Pearl LH, Prodromou C (2000) Structure and in vivo function of Hsp90. Curr Opin Struct Biol 10:46–51

Peng Y, Chen L, Li C, Lu W, Chen J (2001) Inhibition of MDM 2 by hsp90 contributes to mutant p53 stabilization. J Biol Chem 276:40583–40590

Perdew GH (1988) Association of the Ah receptor with the 90-kDa heat shock protein. J Biol Chem 263:13802–13805

Perdew GH, Hord N, Hollenback CE, Welsh MJ (1993) Localization and characterization of the 86- and 84-kDa heat shock proteins in Hepa 1c1c7 cells. Exp Cell Res 209:350–356

Perdew GH, Whitelaw ML (1991) Evidence that the 90-kDa heat shock protein (HSP90) exists in cytosol in heteromeric complexes containing HSP70 and three other proteins with Mr of 63,000, 56,000, and 50,000. J Biol Chem 266:6708–6713

Pfanner N (2000) Protein sorting: recognizing mitochondrial presequences. Curr Biol 10:R412–R415

Pfund C, Lopez-Hoyo N, Ziegelhoffer T, Schilke BA, Lopez-Buesa P, Walter WA, Wiedmann M, Craig EA (1998) The molecular chaperone Ssb from Saccharomyces cerevisiae is a component of the ribosome-nascent chain complex. EMBO J 17:3981–3989

Picard D (2002) Heat-shock protein 90, a chaperone for folding and regulation. Cell Mol Life Sci 59:1640–1648

Pierpaoli EV, Gisler SM, Christen P (1998) Sequence-specific rates of interaction of target peptides with the molecular chaperones DnaK and DnaJ. Biochemistry 37:16741–16748

Piper PW (2001) The Hsp90 chaperone as a promising drug target. Curr Opin Investig Drugs 2:1606–1610

Pratt WB, Scherrer LC, Hutchison KA, Dalman FC (1992) A model of glucocorticoid receptor unfolding and stabilization by a heat shock protein complex. J Steroid Biochem Mol Biol 41:223–229

Pratt WB, Toft DO (1997) Steroid receptor interactions with heat shock protein and immunophilin chaperones. Endocr Rev 18:306–360

Pratt WB, Toft DO (2003) Regulation of signaling protein function and trafficking by the hsp90/hsp70-based chaperone machinery. Exp Biol Med.(Maywood.) 228:111–133

Privalsky ML (1991) A subpopulation of the v-erb A oncogene protein, a derivative of a thyroid hormone receptor, associates with heat shock protein 90. J Biol Chem 266:1456–1462

Prodromou C, Panaretou B, Chohan S, Siligardi G, O'Brien R, Ladbury JE, Roe SM, Piper PW, Pearl LH (2000) The ATPase cycle of Hsp90 drives a molecular "clamp" via transient dimerization of the N-terminal domains. EMBO J 19:4383–4392

Prodromou C, Roe SM, O'Brien R, Ladbury JE, Piper PW, Pearl LH (1997a) Identification and structural characterization of the ATP/ADP-binding site in the Hsp90 molecular chaperone. Cell 90:65–75

Prodromou C, Roe SM, Piper PW, Pearl LH (1997b) A molecular clamp in the crystal structure of the N-terminal domain of the yeast Hsp90 chaperone. Nat Struct Biol 4:477–482

Prodromou C, Siligardi G, O'Brien R, Woolfson DN, Regan L, Panaretou B, Ladbury JE, Piper PW, Pearl LH (1999) Regulation of Hsp90 ATPase activity by tetratricopeptide repeat (TPR)-domain cochaperones. EMBO J 18:754–762

Queitsch C, Sangster TA, Lindquist S (2002) Hsp90 as a capacitor of phenotypic variation. Nature 417:618–624

Rafestin-Oblin ME, Couette B, Radanyi C, Lombes M, Baulieu EE (1989) Mineralocorticosteroid receptor of the chick intestine. Oligomeric structure and transformation. J Biol Chem 264:9304–9309

Ramsey AJ, Russell LC, Whitt SR, Chinkers M (2000) Overlapping sites of tetratricopeptide repeat protein binding and chaperone activity in heat shock protein 90. J Biol Chem 275:17857–17862

Rebbe NF, Ware J, Bertina RM, Modrich P, Stafford DW (1987) Nucleotide sequence of a cDNA for a member of the human 90-kDa heat-shock protein family. Gene 53:235–245

Richter K, Buchner J (2001) Hsp90: chaperoning signal transduction. J Cell Physiol 188:281–290

Richter K, Muschler P, Hainzl O, Buchner J (2001) Coordinated ATP hydrolysis by the Hsp90 dimer. J Biol Chem 276:33689–33696

Richter K, Muschler P, Hainzl O, Reinstein J, Buchner J (2003) Sti1 is a noncompetitive inhibitor of the Hsp90 ATPase. Binding prevents the N-terminal dimerization reaction during the atpase cycle. J Biol Chem 278:10328–10333

Richter K, Reinstein J, Buchner J (2002) N-terminal residues regulate the catalytic efficiency of the Hsp90 ATPase cycle. J Biol Chem 277:44905–44910

Riggs DL, Roberts PJ, Chirillo SC, Cheung-Flynn J, Prapapanich V, Ratajczak T, Gaber R, Picard D, Smith DF (2003) The Hsp90-binding peptidylprolyl isomerase FKBP52 potentiates glucocorticoid signaling in vivo. EMBO J 22:1158–1167

Roe SM, Prodromou C, O'Brien R, Ladbury JE, Piper PW, Pearl LH (1999) Structural basis for inhibition of the Hsp90 molecular chaperone by the antitumor antibiotics radicicol and geldanamycin. J Med Chem 42:260–266

Rose DW, Wettenhall RE, Kudlicki W, Kramer G, Hardesty B (1987) The 90-kilodalton peptide of the heme-regulated eIF-2 alpha kinase has sequence similarity with the 90-kilodalton heat shock protein. Biochemistry 26:6583–6587

Rudiger S, Buchberger A, Bukau B (1997a) Interaction of Hsp70 chaperones with substrates. Nat Struct Biol 4:342–349

Rudiger S, Germeroth L, Schneider-Mergener J, Bukau B (1997b) Substrate specificity of the DnaK chaperone determined by screening cellulose-bound peptide libraries. EMBO J 16:1501–1507

Rudiger S, Mayer MP, Schneider-Mergener J, Bukau B (2000) Modulation of substrate specificity of the DnaK chaperone by alteration of a hydrophobic arch. J Mol Biol 304:245–251

Rudiger S, Schneider-Mergener J, Bukau B (2001) Its substrate specificity characterizes the DnaJ cochaperone as a scanning factor for the DnaK chaperone. EMBO J 20:1042–1050

Rutherford SL, Lindquist S (1998) Hsp90 as a capacitor for morphological evolution. Nature 396:336–342

Sabbah M, Redeuilh G, Baulieu EE (1989) Subunit composition of the estrogen receptor. Involvement of the hormone-binding domain in the dimeric state. J Biol Chem 264:2397–2400

Sakagami M, Morrison P, Welch WJ (1999) Benzoquinoid ansamycins (herbimycin A and geldanamycin) interfere with the maturation of growth factor receptor tyrosine kinases. Cell Stress Chaperones 4:19-28

Saleh A, Srinivasula SM, Balkir L, Robbins PD, Alnemri ES (2000) Negative regulation of the Apaf-1 apoptosome by Hsp70. Nat Cell Biol 2:476–483

Sanchez ER, Meshinchi S, Schlesinger MJ, Pratt WB (1987) Demonstration that the 90-kilodalton heat shock protein is bound to the glucocorticoid receptor in its 9S nondeoxynucleic acid binding form. Mol Endocrinol 1:908–912

Sanchez ER, Redmond T, Scherrer LC, Bresnick EH, Welsh MJ, Pratt WB (1988) Evidence that the 90-kilodalton heat shock protein is associated with tubulin-containing complexes in L cell cytosol and in intact PtK cells. Mol Endocrinol 2:756–760

Sato S, Fujita N, Tsuruo T (2000) Modulation of Akt kinase activity by binding to Hsp90. Proc Natl Acad Sci 97:10832–10837

Schaiff WT, Hruska KA, Jr., McCourt DW, Green M, Schwartz BD (1992) HLA-DR associates with specific stress proteins and is retained in the endoplasmic reticulum in invariant chain negative cells. J Exp Med 176:657–666

Schatz G, Dobberstein B (1996) Common principles of protein translocation across membranes. Science 271:1519–1526

Scheibel T, Neuhofen S, Weikl T, Mayr C, Reinstein J, Vogel PD, Buchner J (1997) ATP-binding properties of human Hsp90. J Biol Chem 272:18608–18613

Scheibel T, Siegmund HI, Jaenicke R, Ganz P, Lilie H, Buchner J (1999) The charged region of Hsp90 modulates the function of the N-terminal domain. Proc Natl Acad Sci 96:1297–1302

Scheibel T, Weikl T, Buchner J (1998) Two chaperone sites in Hsp90 differing in substrate specificity and ATP dependence. Proc Natl Acad Sci 95:1495–1499

Scherrer LC, Dalman FC, Massa E, Meshinchi S, Pratt WB (1990) Structural and functional reconstitution of the glucocorticoid receptor-hsp90 complex. J Biol Chem 265:21397–21400

Scheufler C, Brinker A, Bourenkov G, Pegoraro S, Moroder L, Bartunik H, Hartl FU, Moarefi I (2000) Structure of TPR domain-peptide complexes: critical elements in the assembly of the Hsp70-Hsp90 multichaperone machine. Cell 101:199–210

Schiene-Fischer C, Habazettl J, Schmid FX, Fischer G (2002) The hsp70 chaperone DnaK is a secondary amide peptide bond cis-trans isomerase. Nat Struct Biol 9:419–424

Schmid D, Baici A, Gehring H, Christen P (1994) Kinetics of molecular chaperone action. Science 263:971–973

Schmid SL, Braell WA, Rothman JE (1985) ATP catalyzes the sequestration of clathrin during enzymatic uncoating. J Biol Chem 260:10057–10062

Schmid SL, Rothman JE (1985) Two classes of binding sites for uncoating protein in clathrin triskelions. J Biol Chem 260:10050–10056

Schneider C, Sepp-Lorenzino L, Nimmesgern E, Ouerfelli O, Danishefsky S, Rosen N, Hartl FU (1996) Pharmacologic shifting of a balance between protein refolding and degradation mediated by Hsp90. Proc Natl Acad Sci 93:14536–14541

Schnur RC, Corman ML, Gallaschun RJ, Cooper BA, Dee MF, Doty JL, Muzzi ML, DiOrio CI, Barbacci EG, Miller PE (1995a) erbB-2 oncogene inhibition by geldanamycin derivatives: synthesis, mechanism of action, and structure-activity relationships. J Med Chem 38:3813–3820

Schnur RC, Corman ML, Gallaschun RJ, Cooper BA, Dee MF, Doty JL, Muzzi ML, Moyer JD, DiOrio CI, Barbacci EG (1995b) Inhibition of the oncogene product p185erbB-2 in vitro and in vivo by geldanamycin and dihydrogeldanamycin derivatives. J Med Chem 38:3806–3812

Scholz G, Hartson SD, Cartledge K, Hall N, Shao J, Dunn AR, Matts RL (2000) p50(Cdc37) can buffer the temperature-sensitive properties of a mutant of Hck. Mol Cell Biol 20:6984–6995

Schonfeld HJ, Schmidt D, Schroder H, Bukau B (1995) The DnaK chaperone system of Escherichia coli: quaternary structures and interactions of the DnaK and GrpE components. J Biol Chem 270:2183–2189

Schroder H, Langer T, Hartl FU, Bukau B (1993) DnaK, DnaJ and GrpE form a cellular chaperone machinery capable of repairing heat-induced protein damage. EMBO J 12:4137–4144

Schuh S, Yonemoto W, Brugge J, Bauer VJ, Riehl RM, Sullivan WP, Toft DO (1985) A 90,000-dalton binding protein common to both steroid receptors and the Rous sarcoma virus transforming protein, pp60v-src. J Biol Chem 260:14292–14296

Schulte TW, Blagosklonny MV, Ingui C, Neckers L (1995) Disruption of the Raf-1-Hsp90 molecular complex results in destabilization of Raf-1 and loss of Raf-1-Ras association. J Biol Chem 270:24585–24588

Schumacher RJ, Hurst R, Sullivan WP, McMahon NJ, Toft DO, Matts RL (1994) ATP-dependent chaperoning activity of reticulocyte lysate. J Biol Chem 269:9493–9499

Schwimmer C, Masison DC (2002) Antagonistic interactions between yeast [PSI(+)] and [URE3] prions and curing of [URE3] by Hsp70 protein chaperone Ssa1p but not by Ssa2p. Mol Cell Biol 22:3590–3598

Sepehrnia B, Paz IB, Dasgupta G, Momand J (1996) Heat shock protein 84 forms a complex with mutant p53 protein predominantly within a cytoplasmic compartment of the cell. J Biol Chem 271:15084–15090

Shao J, Grammatikakis N, Scroggins BT, Uma S, Huang W, Chen JJ, Hartson SD, Matts RL (2001) Hsp90 regulates p50(cdc37) function during the biogenesis of the active conformation of the heme-regulated eIF2 alpha kinase. J Biol Chem 276:206–214

Shiu RP, Pouyssegur J, Pastan I (1977) Glucose depletion accounts for the induction of two transformation-sensitive membrane proteins in Rous sarcoma virus-transformed chick embryo fibroblasts. Proc Natl Acad Sci 74:3840–3844

Siegers K, Waldmann T, Leroux MR, Grein K, Shevchenko A, Schiebel E, Hartl FU (1999) Compartmentation of protein folding in vivo: sequestration of nonnative polypeptide by the chaperonin-GimC system. EMBO J 18:75–84

Sikorski RS, Boguski MS, Goebl M, Hieter P (1990) A repeating amino acid motif in CDC23 defines a family of proteins and a new relationship among genes required for mitosis and RNA synthesis. Cell 60:307–317

Siligardi G, Panaretou B, Meyer P, Singh S, Woolfson DN, Piper PW, Pearl LH, Prodromou C (2002) Regulation of Hsp90 ATPase activity by the cochaperone Cdc37p/p50cdc37. J Biol Chem 277:20151–20159

Silverstein AM, Grammatikakis N, Cochran BH, Chinkers M, Pratt WB (1998) p50(cdc37) binds directly to the catalytic domain of Raf as well as to a site on hsp90 that is topologically adjacent to the tetratricopeptide repeat binding site. J Biol Chem 273:20090–20095

Smith DF (1993) Dynamics of heat shock protein 90-progesterone receptor binding and the disactivation loop model for steroid receptor complexes. Mol Endocrinol 7:1418–1429

Smith DF, Faber LE, Toft DO (1990) Purification of unactivated progesterone receptor and identification of novel receptor-associated proteins. J Biol Chem 265:3996–4003

Smith DF, Stensgard BA, Welch WJ, Toft DO (1992) Assembly of progesterone receptor with heat shock proteins and receptor activation are ATP mediated events. J Biol Chem 267:1350–1356

Smith DF, Sullivan WP, Marion TN, Zaitsu K, Madden B, McCormick DJ, Toft DO (1993) Identification of a 60-kilodalton stress-related protein, p60, which interacts with hsp90 and hsp70. Mol Cell Biol 13:869–876

Smith DF, Toft DO (1993) Steroid receptors and their associated proteins. Mol Endocrinol 7:4–11

Smith DF, Whitesell L, Nair SC, Chen S, Prapapanich V, Rimerman RA (1995) Progesterone receptor structure and function altered by geldanamycin, an hsp90-binding agent. Mol Cell Biol 15:6804–6812

Sondermann H, Ho AK, Listenberger LL, Siegers K, Moarefi I, Wente SR, Hartl FU, Young JC (2002) Prediction of novel Bag-1 homologs based on structure/function analysis identifies Snl1p as an Hsp70 cochaperone in Saccharomyces cerevisiae. J Biol Chem 277:33220–33227

Song HY, Dunbar JD, Zhang YX, Guo D, Donner DB (1995) Identification of a protein with homology to hsp90 that binds the type 1 tumor necrosis factor receptor. J Biol Chem 270:3574–3581

Song J, Takeda M, Morimoto RI (2001) Bag1-Hsp70 mediates a physiological stress signaling pathway that regulates Raf-1/ERK and cell growth. Nat Cell Biol 3:276–282

Sorger PK, Pelham HR (1987) The glucose-regulated protein grp94 is related to heat shock protein hsp90. J Mol Biol 194:341–344

Soti C, Racz A, Csermely P (2002) A Nucleotide-dependent molecular switch controls ATP binding at the C-terminal domain of Hsp90. N-terminal nucleotide binding unmasks a C-terminal binding pocket. J Biol Chem 277:7066–7075

Spence J, Cegielska A, Georgopoulos C (1990) Role of Escherichia coli heat shock proteins DnaK and HtpG (C62.5) in response to nutritional deprivation. J Bacteriol 172:7157–7166

Spence J, Georgopoulos C (1989) Purification and properties of the Escherichia coli heat shock protein, HtpG. J Biol Chem 264:4398–4403

Sreedhar AS, Mihaly K, Pato B, Schnaider T, Stetak A, Kis-Petik K, Fidy J, Simonics T, Maraz A, Csermely P (2003) Hsp90 inhibition accelerates cell lysis. Anti-Hsp90 ribozyme reveals a complex mechanism of Hsp90 inhibitors involving both superoxide- and Hsp90-dependent events. J Biol Chem 278:35231–35240

Stancato LF, Chow YH, Hutchison KA, Perdew GH, Jove R, Pratt WB (1993) Raf exists in a native heterocomplex with hsp90 and p50 that can be reconstituted in a cell-free system. J Biol Chem 268:21711–21716
Stancato LF, Silverstein AM, Owens-Grillo JK, Chow YH, Jove R, Pratt WB (1997) The hsp90-binding antibiotic geldanamycin decreases Raf levels and epidermal growth factor signaling without disrupting formation of signaling complexes or reducing the specific enzymatic activity of Raf kinase. J Biol Chem 272:4013–4020
Stebbins CE, Russo AA, Schneider C, Rosen N, Hartl FU, Pavletich NP (1997) Crystal structure of an Hsp90-geldanamycin complex: targeting of a protein chaperone by an antitumor agent. Cell 89:239–250
Stepanova L, Leng X, Parker SB, Harper JW (1996) Mammalian p50Cdc37 is a protein kinase-targeting subunit of Hsp90 that binds and stabilizes Cdk4. Genes Dev 10:1491–1502
Stewart S, Sundaram M, Zhang Y, Lee J, Han M, Guan KL (1999) Kinase suppressor of Ras forms a multiprotein signaling complex and modulates MEK localization. Mol Cell Biol 19:5523–5534
Sullivan WP, Owen BA, Toft DO (2002) The influence of ATP and p23 on the conformation of hsp90. J Biol Chem 277:45942–45948
Szabo A, Langer T, Schroder H, Flanagan J, Bukau B, Hartl FU (1994) The ATP hydrolysis-dependent reaction cycle of the Escherichia coli Hsp70 system DnaK, DnaJ, and GrpE. Proc Natl Acad Sci 91:10345–10349
Szyszka R, Kramer G, Hardesty B (1989) The phosphorylation state of the reticulocyte 90-kDa heat shock protein affects its ability to increase phosphorylation of peptide initiation factor 2 alpha subunit by the heme-sensitive kinase. Biochemistry 28:1435–1438
Takata Y, Imamura T, Iwata M, Usui I, Haruta T, Nandachi N, Ishiki M, Sasaoka T, Kobayashi M (1997) Functional importance of heat shock protein 90 associated with insulin receptor on insulin-stimulated mitogenesis. Biochem Biophys Res Commun 237:345–347
Taniguchi M, Uehara Y, Matsuyama M, Takahashi M (1993) Inhibition of ret tyrosine kinase activity by herbimycin A. Biochem Biophys Res Commun 195:208–214
Teter SA, Houry WA, Ang D, Tradler T, Rockabrand D, Fischer G, Blum P, Georgopoulos C, Hartl FU (1999) Polypeptide flux through bacterial Hsp70: DnaK cooperates with trigger factor in chaperoning nascent chains. Cell 97:755–765
Theyssen H, Schuster HP, Packschies L, Bukau B, Reinstein J (1996) The second step of ATP binding to DnaK induces peptide release. J Mol Biol 263:657–670
Thulasiraman V, Yang CF, Frydman J (1999) In vivo newly translated polypeptides are sequestered in a protected folding environment. EMBO J 18:85–95
Tilly K, McKittrick N, Zylicz M, Georgopoulos C (1983) The dnaK protein modulates the heat-shock response of Escherichia coli. Cell 34:641–646
Toft D, Gorski J (1966) A receptor molecule for estrogens: isolation from the rat uterus and preliminary characterization. Proc Natl Acad Sci 55:1574–1581
Tsubuki S, Saito Y, Kawashima S (1994) Purification and characterization of an endogenous inhibitor specific to the Z-Leu-Leu-Leu-MCA degrading activity in proteasome and its identification as heat-shock protein 90. FEBS Lett 344:229–233
Uzawa M, Grams J, Madden B, Toft D, Salisbury JL (1995) Identification of a complex between centrin and heat shock proteins in CSF-arrested Xenopus oocytes and dissociation of the complex following oocyte activation. Dev Biol 171:51–59
Vaiskunaite R, Kozasa T, Voyno-Yasenetskaya TA (2001) Interaction between the G alpha subunit of heterotrimeric G(12) protein and Hsp90 is required for G alpha(12) signaling. J Biol Chem 276:46088–46093
van den Berg B, Ellis RJ, Dobson CM (1999) Effects of macromolecular crowding on protein folding and aggregation. EMBO J 18:6927–6933
Veldscholte J, Berrevoets CA, Zegers ND, van der Kwast TH, Grootegoed JA, Mulder E (1992) Hormone-induced dissociation of the androgen receptor-heat-shock protein complex: use of a new monoclonal antibody to distinguish transformed from nontransformed receptors. Biochemistry 31:7422–7430
Venema RC, Venema VJ, Ju H, Harris MB, Snead C, Jilling T, Dimitropoulou C, Maragoudakis ME, Catravas JD (2003) Novel complexes of guanylate cyclase with heat shock protein 90 and nitric oxide synthase. Am J Physiol Heart Circ Physiol 285:H669–H678
Vidal V, Qiu NH, Redfield B, Carlino A, Brot N, Weissbach H (1996) ATP hydrolysis is not required for the dissociation of a substance P BiP complex. Arch Biochem Biophys 330:314–318
Voisine C, Cheng YC, Ohlson M, Schilke B, Hoff K, Beinert H, Marszalek J, Craig EA (2001) Jac1, a mitochondrial J-type chaperone, is involved in the biogenesis of Fe/S clusters in Saccharomyces cerevisiae. Proc Natl Acad Sci 98:1483–1488

Wagner BJ, Margolis JW (1995) Age-dependent association of isolated bovine lens multicatalytic proteinase complex (proteasome) with heat-shock protein 90, an endogenous inhibitor. Arch Biochem Biophys 323:455–462

Waldron C, Jund R, Lacroute F (1974) The elongation rate of proteins of different molecular weight classes in yeast. FEBS Lett 46:11–16

Walter S, Buchner J (2002) Molecular chaperones–cellular machines for protein folding. Angew Chem Int Ed Engl 41:1098–1113

Warth R, Briand PA, Picard D (1997) Functional analysis of the yeast 40-kDa cyclophilin Cyp40 and its role for viability and steroid receptor regulation. Biol Chem 378:381–391

Wartmann M, Davis RJ (1994) The native structure of the activated Raf protein kinase is a membrane-bound multi-subunit complex. J Biol Chem 269:6695–6701

Weaver AJ, Sullivan WP, Felts SJ, Owen BA, Toft DO (2000) Crystal structure and activity of human p23, a heat shock protein 90 cochaperone. J Biol Chem 275:23045–23052

Wegele H, Haslbeck M, Reinstein J, Buchner J (2003a) Sti1 is a novel activator of the Ssa proteins. J Biol Chem 278:25970–25976

Wegele H, Muschler P, Bunck M, Reinstein J, Buchner J (2003b) Dissection of the contribution of individual domains to the ATPase mechanism of Hsp90. J Biol Chem 278:39303–39310

Wei J, Gaut JR, Hendershot LM (1995) In vitro dissociation of BiP-peptide complexes requires a conformational change in BiP after ATP binding but does not require ATP hydrolysis. J Biol Chem 270:26677–26682

Weikl T, Abelmann K, Buchner J (1999) An unstructured C-terminal region of the Hsp90 cochaperone p23 is important for its chaperone function. J Mol Biol 293:685–691

Weikl T, Muschler P, Richter K, Veit T, Reinstein J, Buchner J (2000) C-terminal regions of Hsp90 are important for trapping the nucleotide during the ATPase cycle. J Mol Biol 303:583–592

Welch WJ, Feramisco JR (1982) Purification of the major mammalian heat shock proteins. J Biol Chem 257:14949–14959

Werner-Washburne M, Craig EA (1989) Expression of members of the Saccharomyces cerevisiae hsp70 multigene family. Genome 31:684–689

Whitelaw ML, Hutchison K, Perdew GH (1991) A 50-kDa cytosolic protein complexed with the 90-kDa heat shock protein (hsp90) is the same protein complexed with pp60v-src hsp90 in cells transformed by the Rous sarcoma virus. J Biol Chem 266:16436–16440

Whitesell L, Mimnaugh EG, De Costa B, Myers CE, Neckers LM (1994) Inhibition of heat shock protein HSP90-pp60v-src heteroprotein complex formation by benzoquinone ansamycins: essential role for stress proteins in oncogenic transformation. Proc Natl Acad Sci 91:8324–8328

Whitesell L, Shifrin SD, Schwab G, Neckers LM (1992) Benzoquinonoid ansamycins possess selective tumoricidal activity unrelated to src kinase inhibition. Cancer Res 52:1721–1728

Whitesell L, Sutphin PD, Pulcini EJ, Martinez JD, Cook PH (1998) The physical association of multiple molecular chaperone proteins with mutant p53 is altered by geldanamycin, an hsp90-binding agent. Mol Cell Biol 18:1517–1524

Wickner S, Hoskins J, McKenney K (1991) Function of DnaJ and DnaK as chaperones in origin-specific DNA binding by RepA. Nature 350:165–167

Wigley DB, Davies GJ, Dodson EJ, Maxwell A, Dodson G (1991) Crystal structure of an N-terminal fragment of the DNA gyrase B protein. Nature 351:624–629

Wilbanks SM, Chen L, Tsuruta H, Hodgson KO, McKay DB (1995) Solution small-angle X-ray scattering study of the molecular chaperone Hsc70 and its subfragments. Biochemistry 34:12095–12106

Wilbanks SM, McKay DB (1995) How potassium affects the activity of the molecular chaperone Hsc70. II. Potassium binds specifically in the ATPase active site. J Biol Chem 270:2251–2257

Wilhelmsson A, Cuthill S, Denis M, Wikstrom AC, Gustafsson JA, Poellinger L (1990) The specific DNA binding activity of the dioxin receptor is modulated by the 90 kd heat shock protein. EMBO J 9:69–76

Wilson EM, Lea OA, French FS (1977) 9S binding protein for androgens and progesterone. Proc Natl Acad Sci 74:1960–1964

Xu M, Dittmar KD, Giannoukos G, Pratt WB, Simons SS, Jr (1998) Binding of hsp90 to the glucocorticoid receptor requires a specific 7-amino acid sequence at the amino terminus of the hormone-binding domain. J Biol Chem 273:13918–13924

Xu W, Mimnaugh E, Rosser MF, Nicchitta C, Marcu M, Yarden Y, Neckers L (2001) Sensitivity of mature Erbb2 to geldanamycin is conferred by its kinase domain and is mediated by the chaperone protein Hsp90. J Biol Chem 276:3702–3708

Xu Y, Lindquist S (1993) Heat-shock protein hsp90 governs the activity of pp60v-src kinase. Proc Natl Acad Sci 90:7074–7078

Xu Y, Singer MA, Lindquist S (1999) Maturation of the tyrosine kinase c-src as a kinase and as a substrate depends on the molecular chaperone Hsp90. Proc Natl Acad Sci 96:109–114
Young JC, Hartl FU (2000) Polypeptide release by Hsp90 involves ATP hydrolysis and is enhanced by the cochaperone p23. EMBO J 19:5930–5940
Young JC, Moarefi I, Hartl FU (2001) Hsp90: a specialized but essential protein-folding tool. J Cell Biol 154:267–273
Young JC, Obermann WM, Hartl FU (1998) Specific binding of tetratricopeptide repeat proteins to the C-terminal 12-kDa domain of hsp90. J Biol Chem 273:18007–18010
Young JC, Schneider C, Hartl FU (1997) In vitro evidence that hsp90 contains two independent chaperone sites. FEBS Lett 418:139–143
Yura T, Nagai H, Mori H (1993) Regulation of the heat-shock response in bacteria. Annu Rev Microbiol 47:321–350
Zhang L, Hach A, Wang C (1998) Molecular mechanism governing heme signaling in yeast: a higher-order complex mediates heme regulation of the transcriptional activator HAP1. Mol Cell Biol 18:3819–3828
Zhao YG, Gilmore R, Leone G, Coffey MC, Weber B, Lee PW (2001) Hsp90 phosphorylation is linked to its chaperoning function. Assembly of the reovirus cell attachment protein. J Biol Chem 276:32822–32827
Zhu X, Zhao X, Burkholder WF, Gragerov A, Ogata CM, Gottesman ME, Hendrickson WA (1996) Structural analysis of substrate binding by the molecular chaperone DnaK. Science 272:1606–1614
Ziemiecki A, Catelli MG, Joab I, Moncharmont B (1986) Association of the heat shock protein hsp90 with steroid hormone receptors and tyrosine kinase oncogene products. Biochem Biophys Res Commun 138:1298–1307
Zylicz M, Ang D, Georgopoulos C (1987) The grpE protein of Escherichia coli. Purification and properties. J Biol Chem 262:17437–17442
Zylicz M, King FW, Wawrzynow A (2001) Hsp70 interactions with the p53 tumor suppressor protein. EMBO J 20:4634–4638

Rev Physiol Biochem Pharmacol (2004) 151:45–91
DOI 10.1007/s10254-004-0022-8

R. Schülein

The early stages of the intracellular transport of membrane proteins: clinical and pharmacological implications

Published online: 23 April 2004

Abstract Intracellular transport mechanisms ensure that integral membrane proteins are delivered to their correct subcellular compartments. Efficient intracellular transport is a prerequisite for the establishment of both cell architecture and function. In the past decade, transport processes of proteins have also drawn the attention of clinicians and pharmacologists since many diseases have been shown to be caused by transport-deficient proteins. Membrane proteins residing within the plasma membrane are transported via the secretory (exocytotic) pathway. The general transport routes of the secretory pathway are well established. The transport of membrane proteins starts with their integration into the ER membrane. The ribosomes synthesizing membrane proteins are targeted to the ER membrane, and the nascent chains are co-translationally integrated into the bilayer, i.e., they are inserted while their synthesis is in progress. During ER insertion, the orientation (topology) of the proteins in the membrane is determined. Proteins are folded, and their folding state is checked by a quality control system that allows only correctly folded forms to leave the ER. Misfolded or incompletely folded forms are retained, transported back to the cytosol and finally subjected to proteolysis. Correctly folded proteins are transported in the membranes of vesicles through the ER/Golgi intermediate compartment (ERGIC) and the individual compartments of the Golgi apparatus (*cis*, *medial*, *trans*) to the plasma membrane. In this review, the current knowledge of the first stages of the intracellular trafficking of membrane proteins will be summarized. This "early secretory pathway" includes the processes of ER insertion, topology determination, folding, quality control and the transport to the Golgi apparatus. Mutations in the genes of membrane proteins frequently lead to misfolded forms that are recognized and retained by the quality control system. Such mutations may cause inherited diseases like cystic fibrosis or retinitis pigmentosa. In the second part of this review, the clinical implications of the early secretory pathway will be discussed. Finally, new pharmacological strategies to rescue misfolded and transport-defective membrane proteins will be outlined.

R. Schülein (✉)
Forschungsinstitut für Molekulare Pharmakologie (FMP), Robert-Rössle-Str. 10, 13125 Berlin, Germany
e-mail: schuelein@fmp-berlin.de · Tel.: +49-30-94793255 · Fax: +49-30-94793109

Abbreviations *AP1:* Clathrin-associated adaptor protein complex 1 · *AQP:* Aquaporin · *ARF:* ADP-ribosylation factor · *AVP:* 8-Arginine-vasopressin;BiP immunoglobulin heavy chain binding protein · *CFTR:* Cystic fibrosis transmembrane conductance regulator · *CLQTS:* Congenital long QT syndrome · *CMT:* Charcot-Marie-Tooth syndrome · *CNX:* Calnexin · *COPI:* Coat protein complex I · *COPII:* Coat protein complex II · *CPX:* 8-Cyclopentyl-1,2-dipropylxanthine · *CRT:* Calreticulin · *CSID:* Congenital sucrose-isomaltase deficiency · *Cx:* Connexin · *cGMP:* Cyclic 3′:5′ guanosine monophosphate · *ECL:* Extracellular loop · *EndoH:* Endoglycosidase H · *ER:* Endoplasmic reticulum · *ERAD:* ER-associated degradation · *ERGIC:* ER/Golgi intermediate compartment · *ERp:* ER protein · ET_BR*:* Human endothelin B receptor · *FH:* Familial hypercholesterolemia · *GABA:* Gamma amino butyric acid · *GFP:* Green fluorescent protein · *GH:* Growth hormone · *GHIS:* Growth hormone insensitivity syndrome · *GLCase:* Glucosidase · *GlcNac:* *N*-acetylglucosamine · *GPCR:* G protein-coupled receptor · *GPI:* Glycosylphosphatidylinositol · *G protein:* GTP-binding protein · *GRP:* Glucose-regulated protein · *HA:* Hemagglutinin · *Hdj-2:* Human DnaJ-2 protein · *HFE:* Human hemochromatosis protein · *HH:* Hereditary hemochromatosis · *HEK 293 cells:* Human embryonic kidney 293 cells · *HERG:* Human ether-a-go-go-related protein · *Hsc70:* Heat shock cognate 70 protein · *ICL:* Intracellular loop · *IGF-I:* Insulin-like growth factor-1 · I_{Kr}: Rapidly activating delayed rectifier potassium current · I_{Ks}: Slowly activating delayed rectifier potassium current · *JAK:* Janus kinase · *LDL:* Low-density lipoprotein · *LH:* Luteinizing hormone/choriogonadotropin · *LS:* Laron syndrome · *MATP:* Membrane associated transporter protein · *MDCK cells:* Madin-Darby canine kidney epithelial cells · *MHC:* Major histocompatibility complex · *MiRP1:* minK-related peptide 1 · *NDI:* Congenital nephrogenic diabetes insipidus · *NMDA:* *N*-methyl-*d*-aspartate · *OCA:* Oculocutaneous albinism · *PDI:* Protein disulfide isomerase · *Pgp:* P-glycoprotein · *PKA:* Protein kinase A · *PLP:* Proteolipid protein · *PMP22:* Peripheral myelin protein 22 · *RP:* Primary retinitis pigmentosa · *SI:* Sucrase-isomaltase · *SNARE:* Ethylmaleimide-sensitive factor attachment protein · *SRP:* Signal recognition particle · *TCR:* T-cell antigen receptor · *TM:* Transmembrane domain · *TRAM:* Translocating chain-associated membrane protein · *Tyr:* Tyrosinase · *Tyrp1:* Tyrosinase-related protein-1 · *UGGT:* UDP-glucose:glycoprotein glucosyltransferase · *VIP:* Vesicular-integral membrane protein · V_2R*:* Vasopressin V_2 receptor · *VSV:* Vesicular stomatitis virus

ER targeting and insertion of membrane proteins

The basic mechanism

The first step of the intracellular transport of membrane proteins is their insertion into the ER membrane. ER targeting/insertion is mediated by signal sequences contained in the proteins and by a complex targeting/insertion machinery (Brodsky 1998). The mechanisms of protein translocation at the ER membrane were studied mainly for secretory proteins which are not integrated into but translocated across the membrane. However, it is generally accepted that membrane proteins use the same targeting and insertion mechanisms (Chin et al. 2002). The following section will summarize the basic mechanisms of membrane protein integration into the ER membrane.

Membrane proteins are initially synthesized at cytoplasmic ribosomes. Cytoplasmic translation stops ("elongation arrest"; Walter and Blobel 1981) or is delayed when the first hydrophobic segment of the membrane protein, the signal sequence, appears. The elongation arrest is mediated by the binding of the signal recognition particle (SRP), an RNA-protein complex. The nascent chain/ribosome/SRP complex is then targeted to the SRP receptor at the ER membrane. The nascent chain is transferred in a GTP-dependent manner to a protein-conducting translocase complex, consisting mainly of the Sec61 channel protein subunit (Görlich et al. 1992; Sec61p in Yeast, Sec61α in mammals; SecY in bacteria and archae) and the translocating chain-associated membrane protein (TRAM; Görlich and Rapoport 1993). After SRP release, translation starts again (Gilmore and Blobel 1983), and the emerging protein is processed co-translationally into the Sec61p channel and from there into the bilayer.

Extensions and variations of the basic mechanism

Recent results demonstrate that the basic mechanism of membrane protein biogenesis at the ER membrane is more complex than originally thought. For example, variations in the ER targeting/insertion mechanisms of membrane proteins are due to the fact that membrane proteins occur in different topologies and use different signal sequences (see the section entitled "ER targeting and insertion of membrane proteins with different topologies and signal sequences"). Furthermore, analyzing the processes of membrane protein topology determination (see "Establishment of correct membrane protein topology") and of how membrane proteins leave the translocase complex channel (see "Leaving the channel: membrane protein integration into the ER membrane") led to a much more detailed view of the mechanism outlined above.

ER targeting and insertion of membrane proteins with different topologies and signal sequences

The most obvious variations of the basic mechanism are due to the fact that membrane proteins possess different topologies and that each type uses a different type of signal sequence for ER targeting. According to the classification of Goder and Spiess (2001), three different types of membrane proteins are distinguished (Fig. 1). Type I membrane proteins contain cleavable signal peptides for ER targeting/insertion which are removed during or

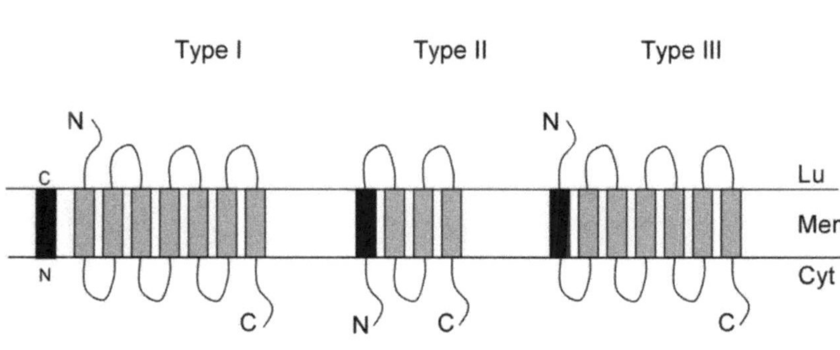

Fig. 1 Different types of multispanning membrane proteins. The classification was adapted from Goder and Spiess (2001). The signal sequence (cleavable signal peptide or non-cleaved signal anchor sequence) is indicated in black. *Lu* ER lumen, *Mem* ER membrane, *Cyt* cytoplasm

after the integration process by the signal peptidases of the ER. The N tails of type I membrane proteins are translocated across the ER membrane, i.e., they are finally located extracellularly (or in the lumen of intracellular membrane compartments in the case of intracellular proteins). Type II proteins do not contain cleavable signal peptides. ER targeting is instead mediated by the first transmembrane domains of the mature proteins that remain in the proteins as non-cleaved signal anchor sequences. The N tails of these proteins are not translocated across the ER membrane, i.e., they are finally located intracellularly. As a consequence, the signal anchor sequences are orientated with the N terminus toward the cytoplasm. In type III proteins, ER targeting is also mediated by non-cleaved signal anchor sequences. The N tails of these proteins are, however, translocated across the ER membrane to a final extracellular or luminal location. The signal anchor sequences of these proteins are consequently orientated with the C terminus toward the cytoplasm. Type I and type III proteins share the same topology with an extracellular N tail. The reason that type I proteins require additional cleavable signal peptides for ER insertion was not known.

ER targeting and insertion of type I and type III proteins: G protein-coupled receptors (GPCRs) as model proteins to study the significance of cleavable signal peptides

The heptahelical GPCRs were ideally suited to answer this question since both types of signal sequences are present in this huge protein family (Wallin and von Heijne 1995). The majority of the receptors (90–95%) are type III proteins containing noncleaved signal anchor sequences. A smaller group (5–10%) contains additional cleavable signal peptides.

A hypothesis for the significance of cleavable signal peptides of type I proteins was developed recently (Fig. 2). As described above, membrane proteins are normally integrated co-translationally into the ER membrane by membrane-bound ribosomes. The initial cytosolic translation continues until the first hydrophobic segment appears, a cleavable signal peptide in the case of a type I protein, a non-cleaved signal anchor sequence (TM1 of the mature protein) in the case of a type III protein. Appearance of the signal leads in either case to the elongation arrest by SRP binding, which lasts until targeting to the transloca-

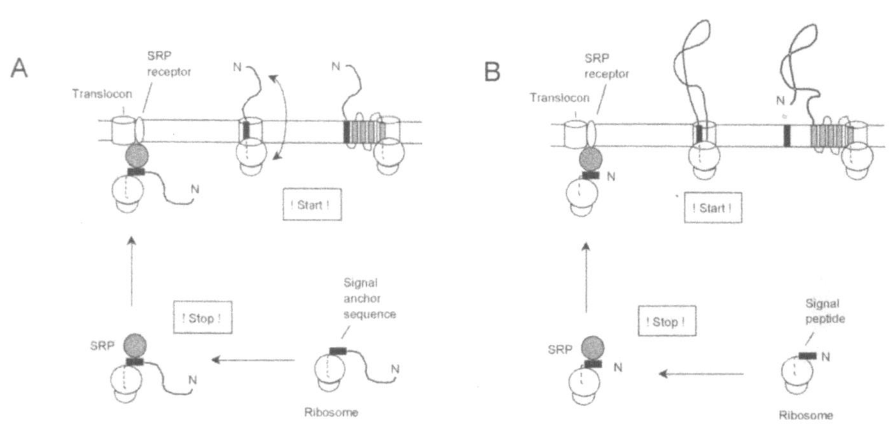

Fig. 2 Putative ER targeting and insertion mechanisms of GPCRs without (**A**) and with (**B**) signal peptide. The respective ER targeting signals (internal signal anchor sequence or cleavable signal peptides) are indicated by *black boxes*. *SRP* signal recognition particle. See text for details. (Adapted from Köchl et al. 2002)

tion complex is complete. This mechanism implies major differences in the translocation of the N termini of type I and type III proteins. For a type III protein without a signal peptide (Fig. 2A), the N terminus is completely synthesized in the cytoplasm, since translation is not stopped until the signal anchor sequence (TM1) appears. In this case, the N terminus must be translocated post-translationally through the Sec61 channel across the ER membrane. In contrast, for a type I protein containing an additional signal peptide (Fig. 2B), the N terminus is not translated in the cytosol, since SRP binding to the preceding signal peptide stops elongation. Here, the N terminus can be translocated co-translationally through the translocase complex. Taking these considerations into account, it was logical to speculate that signal peptides are necessary for those membrane proteins for which a post-translational translocation of the N terminus is impaired. Consistent with this view, it was shown that the signal peptide mutant of the human endothelin B receptor (ET_BR), a GPCR, was unable to translocate the N tail across the ER membrane (Köchl et al. 2002).

What are the N tail properties that impair a post-translational translocation across the ER membrane? For the ET_BR, it was speculated that the presence of a folded domain necessitates a cleavable signal peptide (Köchl et al. 2002). Consistent with this hypothesis, signal peptides are preferentially found in those GPCR families for which a contribution of the N tail to the ligand-binding domain has been demonstrated or proposed (Köchl et al. 2002). For these receptors, it is very likely that they contain stably folded domains in their N tails. Another hypothesis for the requirement of cleavable signal peptides is based on a statistical analysis of GPCR sequences (Wallin and von Heijne 1995). It was proposed that signal peptides are necessary if the N tails exceed a critical length. In line with this notion, it was recently shown that shortening the N tail of the cannaboid receptor 1 increases the amount of correctly folded receptor in the ER membrane (Andersson et al. 2003). The folding and length hypotheses do not exclude each other. The probability of the presence of stably folded domains perhaps increases with N tail length.

The results discussed above do not necessarily mean that all type I proteins are dependent on their signal peptides for their function. For example, the rat brain Na^+-Ca^{2+} exchanger (Furman et al. 1995), the human renal Na^+-Ca^{2+} exchanger (Loo et al. 1995), and the human UDP-glucuronosyltransferase 1A6 (Ouzzine et al. 1999a; Ouzzine et al. 1999b) were processed correctly in the absence of their signal peptides. Chaperones may prevent N tail folding in the cytoplasm in these cases, thereby maintaining sequences in an unfolded and translocation-competent state.

In addition, the results do not allow the conclusion that the type III proteins containing only signal anchor sequences must necessarily have unfolded N tails. The recently published crystal structure of rhodopsin (the archetypal GPCR without signal peptide) revealed that folded domains are present in the N tail (Palczewski et al. 2000; Okada and Palczewski 2001). Chaperones may also be effective in this case. Signal peptides may, however, become a necessity when folding is particularly rapid and stable, or when a critical number of folded domains is exceeded.

The exact mechanism by which the N tails of type III proteins are translocated across the ER membrane is unknown. It is conceivable that they are processed through the Sec61 channel. Two mechanisms of how the N tails pass the channel are possible: (a) N tail translocation may precede TM1 integration into the channel, in which case the N tail is translocated in an N-to-C-terminal direction; (b) N tail translocation may occur after TM1 integration into the channel. In this case, the N tail must be translocated in a C-to-N-terminal direction. Recent data obtained for the ProW protein (a heptahelical type III protein with a GPCR topology) indicate that N tail translocation occurs in a C-to-N-terminal direction (Nilsson et al. 2000), consistent with the second mechanism.

ER targeting and insertion mechanisms of type II proteins: use of unusual signal anchor sequences

For membrane proteins without cleavable signal peptides (type II and type III), the original concept of membrane protein integration into the ER membrane implies that this process is mediated by TM1 (Wessels and Spiess 1988; see also Fig. 2). Recent studies with the cystic fibrosis transmembrane conductance regulator (CFTR), a type II protein, indicate that this concept must be extended. The CFTR protein is integrated into the ER membrane by two different mechanisms (Lu et al. 1998). According to the first mechanism, TM1 mediates ER insertion as a classical signal anchor sequence. According to the second mechanism, however, ER membrane integration is mediated by TM2. TM1 is embedded post-translationally afterward. The second mechanism also applies to the Shaker-like potassium channel (Tu et al. 2000). However, in the case of this protein, TM1 is unable to initiate ER insertion, and this process is mediated by TM2 alone.

It is not known whether the second mechanism is also used by the type III proteins containing extracellular N tails. The heptahelical GPCRs again appear to represent a good model to address this question. For the human vasopressin V2 receptor (V_2R), it was shown that TM1 mediates ER integration (Schülein et al. 1996a). The other TMs of the V_2R, however, also possess a strong potential to function as signal anchor sequences. This can be concluded from studies with receptor truncations, placing the internal TMs into the TM1 position. A receptor fragment comprising only the sequence from TM6 to TM7, for example, was integration-competent (Schöneberg et al. 1996). Similar results were obtained for internal TMs of the m3 muscarinic receptor (Schöneberg et al. 1995). It is not known, however, whether the internal TMs of GPCRs also function as signal anchor sequences when the normal TM1 is present, as is the case for the CFTR protein.

In summary, the results discussed above show that the internal transmembrane domains of polytopic membrane proteins play a more active role in the ER insertion process than previously thought.

Establishment of correct membrane protein topology

During ER insertion, the topology of a membrane protein is established. The transmembrane domains are orientated; the extracellular (luminal) loops are translocated across the ER membrane while the cytoplasmic loops remain on the cytoplasmic side. Once the topology is established, it is maintained throughout the secretory pathway. What are the structural features that determine correct membrane protein orientation? Two related hypotheses have been proposed:

a. The "positive inside" rule was originally postulated for membrane proteins of the inner membrane of *Escherichia coli*. The inner membrane also serves as a model for the ER membrane of eucaryotes. The positive inside rule predicts that the orientation of a membrane protein is determined by the asymmetric distribution of positively charged residues in the sequences flanking the TMs, and that an excess of positively charged residues defines a cytoplasmic domain (von Heijne and Gavel 1988). The strong topogenic potential of positively charged residues at the inner membrane has been demonstrated in a large number of studies (von Heijne 1994).

b. The second hypothesis, the "charge difference" rule, was proposed for eucaryotic membrane proteins and the ER membrane (Hartmann et al. 1989). It postulates that the charge differences between the segments flanking the TMs determine orientation, rather than positively charged residues themselves. According to the charge difference rule, negatively and positively charged residues possess an equal topogenic potential. In studies using the yeast pheromone receptor STE2p, a GPCR, the predictions of the charge difference rule and thus the significance of negatively charged residues were confirmed (Harley and Tipper 1996; Harley et al. 1998).

The charge difference rule in eucaryotes raised the question of whether charged residues have to lie within a critical distance to the transmembrane domain to exert their topogenic function. The statistical study of Hartmann et al. (1989; see above) suggests that such a window region exists since the best correlation was obtained when 10–15 residues on each side were taken into account. Recent experimental data indicate at least that negative charges are only functional when they lie within a critical distance of 6 residues of the transmembrane domain, thus proving the existence of such a window region (Rutz et al. 1999).

The exact mechanism by which charged residues influence membrane protein topology was studied at the inner membrane of *E. coli* only. It was shown that the membrane potential (positive outside, negative inside) promotes the translocation of negative charges by an electrophoretic mechanism, while it impedes that of positive charges (Andersson and von Heijne 1994; Cao et al. 1995; Kiefer et al. 1997; Delgado-Partin and Dalbey 1998). In *E. coli*, the head groups of negatively charged membrane phospholipids may also contribute to the retention of positive charges on the cytoplasmic side of the inner membrane (van Klompenburg et al. 1997). It remains to be determined whether a similar mechanism holds true at the ER membrane of eucaryotes. It was speculated that protein components, such as TRAM, associated with the translocation channel may also play a role in determining the topology of membrane proteins of eucaryotes (Spiess 1995). Protein/protein interactions may also explain why some eucaryotic proteins such as the prion protein (PrP) or the myelin proteolipid protein (PLP) can be synthesized in two different orientations (Ott and Lingappa 2002).

Another question is whether polytopic membrane proteins contain only one or more topogenic determinants. In principle, the topology of a full-length, multispanning membrane protein can be established by determining only the orientation of the first transmembrane domain. The others may follow with an opposite orientation to the preceding one (Wessels and Spiess 1988; Hartmann et al. 1989). Alternatively, each or at least some transmembrane domains may possess their own topogenic determinants and adopt correct orientation independently of the others. Recent results indicate that the latter may be the case. Mutations preventing insertion of the first or the second transmembrane domain of the Glut1 glucose transporter into the ER membrane failed to alter the topology of the downstream protein regions (Sato et al. 1998). A similar significance of downstream topogenic determinants was also described for the ProW protein (Nilsson et al. 2000) and the glycine receptor α subunit (Sadtler et al. 2003).

Leaving the channel: membrane protein integration into the ER membrane

The TMs of a membrane protein are inserted into the lumen of the Sec61 translocation channel immediately after their synthesis. Integration into the ER membrane then requires a unique lateral opening reaction of the channel protein (Martoglio et al. 1995; van den Berg 2004). The main questions arising from this process have been: (a) Do transmembrane domains leave the translocation channel while the synthesis of the protein is in progress ("sequential integration"), or (b) do they remain in the translocation channel until the protein is completely synthesized ("integration upon completion")?

In a study using the P-glycoprotein (Pgp), Borel and Simon (1996) reported that translocated TMs can still be extracted from the membrane with urea, concluding that the TMs are not integrated into the bilayer until protein synthesis is complete. A similar conclusion was drawn from a photocrosslinking study. Here, an artificial fusion protein was not able to leave the Sec61/TRAM complex until translation was terminated (Do et al. 1996). However, conflicting data came from a study using the leader peptidase of *E. coli* as a model. Here, the TMs were released into the ER membrane while translation was in progress, consistent with the first model (Mothes et al. 1997).

A unifying model was presented by Heinrich et al. (2000). It implies that the translocase complex establishes several environments with variable hydrophobicity. The newly synthesized TMs of a membrane protein may stay in the very hydrophobic lipid phase in the vicinity of the Sec61 channel, or in the less hydrophobic TRAM/Sec61 interface, or in the hydrophilic Sec61 channel lumen. The preferred compartment is determined by the hydrophobicity of the individual TMs. A highly hydrophobic TM may quickly leave the translocase complex and enter the lipid phase. Such a behavior is consistent with the first model. A moderately hydrophobic TM may stay for a longer time at the Sec61/TRAM interface or in the channel lumen. It may even be associated with the translocase complex when the next TM arrives. Such a behavior is consistent with the second model and may explain some of the crosslinking data and also the proposed cooperation of transmembrane segments during ER membrane integration (Heinrich and Rapoport 2003). However, the original "integration upon completion" hypothesis of the second model, proposing that a membrane protein completely assembles in the lumen of the channel, does not hold true. The recently published X-ray structure of the Sec61 (SecY) complex of the archae *Methanococcus jannaschii* (van den Berg et al. 2004) revealed that the protein-conducting channel is too small to harbor a polytopic membrane protein.

Folding of membrane proteins in the ER membrane and quality control

During or after insertion into the ER membrane, membrane proteins are folded. Folding is checked by a quality control system, and only correctly folded proteins are allowed to leave the early secretory pathway in transport vesicles. Misfolded forms are retained, transported back to the cytosol and finally subjected to proteolysis.

In the luminal part of the ER, protein folding is assisted by different molecular chaperones. Some of them also play an important role in quality control (see below). "Classical" chaperones bind to unfolded proteins. The glucose-regulated proteins GRP78 (also known as immunoglobulin heavy chain binding protein, BiP), GRP94 and GRP170 (Gething 1999) belong to this group. The "nonclassical" chaperones play an important role in the folding process of proteins that contain N glycosylations. Chaperones belonging to this group are the lectine-like chaperones calnexin (CNX) and calreticulin (CRT), the key components of the CNX cycle (Trombetta and Helenius 1998; Parodi 2000; see below). When the proteins contain disulfide bonds, enzymes catalyzing protein disulfide bond formation and isomerization, such as ERp57 (also known as protein disulfide isomerase, PDI), ERp59 and ERp72 are also involved in the folding process (Noiva 1999).

The chaperones of the ER lumen are well characterized. Comparatively little is known about chaperones assisting membrane protein folding at the cytoplasmic side of the ER membrane. In the case of the CFTR protein, it was shown that the heat shock cognate 70 protein (Hsc70), together with the human DnaJ-2 protein (Hdj-2), may exert such a function (Meacham et al. 1999). Hsc70 appears to be also involved in membrane protein degradation since it can be converted by the co-chaperone CHIP from a protein-folding machine into a degradation factor (Meacham et al. 2001). It was proposed that the CHIP/Hsc70 pair recognizes the folding state of the CFTR protein and targets aberrant forms for proteasomal degradation by promoting their ubiquitination (Meacham et al. 2001; Cyr et al. 2002). Thus, the chaperones at the cytoplasmic side of the ER membrane play an important role in quality control.

The function of the quality control system is to ensure that only correctly folded proteins leave the ER. Misfolded forms are retained and finally subjected to ER-associated degradation (ERAD), a process that includes the retranslocation to the cytosol, ubiquitination and degradation by the proteasome (Bonifacino and Weissmann 1998). The crunch questions of quality control are the following: (a) how does the system differentiate correctly folded from misfolded proteins, and (b) how are misfolded proteins retained in the ER? Although we are far from understanding these mechanisms in detail, a lot of progress has been made in the past decade.

It was originally thought that quality control is mediated exclusively by the ER chaperones which recognize their misfolded substrates mainly by nonspecific hydrophobic interactions (see below). Recent results, however, indicate that specific signal-based mechanisms may also play a role. Moreover, the quality control appears not to be restricted to the ER. At least some misfolded proteins reach the ERGIC. Here, they are recognized and delivered to a retrograde transport system which directs them back to the ER. The following sections of this review will discuss function and localization of the individual components of the quality control system.

Chaperone-mediated quality control: the CNX/CRT cycle

Chaperones have two functions. They assist normal protein folding, and they are involved in the retention of misfolded forms. The latter process is accompanied by a prolonged interaction of the chaperones with their target proteins. The mechanisms by which chaperones assess the folding state of proteins are particularly well studied for glycoproteins. Quality control of proteins carrying N glycosylations involves the two homologous lectin-like chaperones CNX and CRT (Trombetta and Helenius 1998; Parodi 2000; Hauri et al. 2000). The N glycosylation of proteins starts in the ER lumen by the transfer of the core oligosaccharide to the nascent protein. The complete core glycan contains three glucose molecules (Fig. 3). Trimming by the glucosidases I and II leads to a monoglucosylated form which is specifically recognized by CNX (membrane-bound) and/or CRT (soluble) and the co-chaperone ERp57. Many glycoproteins interact with both lectine chaperones,

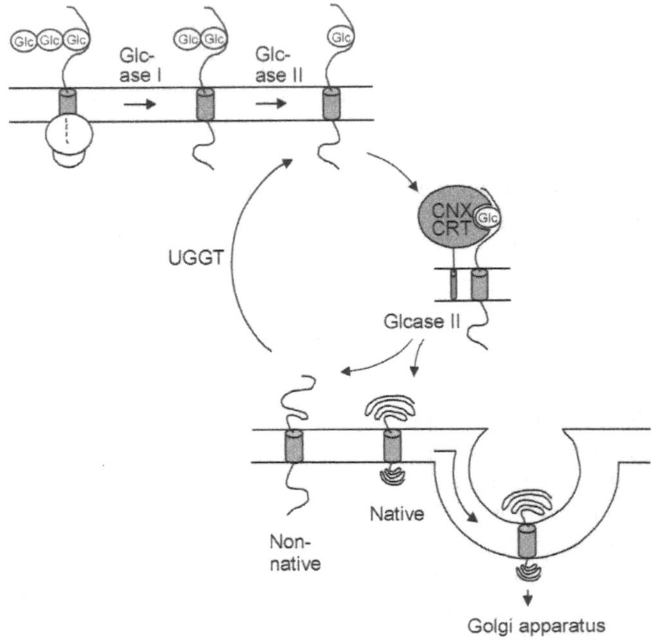

Fig. 3 The CNX cycle. The lectine-like chaperones CNX and CRT bind to non-native monoglucosylated glycoproteins. CNX/CRT binding is terminated by glucosidase II-mediated removal of the single glucose molecule. Proteins carrying non-glucosylated core glycans can leave the ER via the secretory pathway if they have reached their native conformations. Proteins in non-native conformations are reglucosylated by UGGT and enter a novel CNX cycle. *Glc* glucose, *Glc-ase* glucosidase, *CNX* calnexin, *CRT* calreticulin, *UGGT* UDP-glucose:glycoprotein glucosyltransferase. Note that CRX is located in the ER lumen (not depicted)

although the substrate specificities of the two chaperones do not completely overlap (van Leeuwen and Kearse 1996; Danilczyk et al. 2000). Investigation of the exact binding mechanism will be facilitated by the now-available structural data for CNX (Schrag et al. 2001) and CRT (Ellgaard et al. 2001a; Ellgaard et al. 2001b). Binding of CNX and CRT is terminated by the glucosidase II-mediated removal of the last glucose molecule from the glycan. If the resulting nonglucosylated protein has reached its native conformation, it is allowed to leave the ER in vesicles destined for the ERGIC. If the protein has not reached its native conformation, the enzyme UDP-glucose:glycoprotein glucosyltransferase (UGGT) adds a glucose molecule. The resulting monoglucosylated form is again recognized by CNX and CRT, and its ER association is consequently prolonged. Additional deglucosylation, glucosylation, and CNX/CRT binding cycles may be required until the glycoprotein finally reaches its correct folding state.

In the "CNX cycle" outlined above, the enzyme UGGT plays a key role since it senses the folding state of the glycoprotein. A novel glucose molecule is added only to non–native protein substrates (Trombetta and Parodi 1992; Sousa et al. 1992). Whereas it is conceivable that multiple CNX cycles improve the folding efficiency of a glycoprotein, the exact mechanism by which UGGT differentiates correctly folded from non-native forms is still unknown. Recently, models were proposed to explain how this differentiation is achieved (Elgaard and Helenius 2001).

According to the first model, UGGT recognizes non-native glycoproteins by their exposed hydrophobic domains. This model is based on the idea that in correctly folded proteins hydrophobic domains are less likely to be present at the water-exposed protein surface. Such a model is supported by the data of Sousa and Parodi (1995), demonstrating that UGGT binds to hydrophobic peptides in vitro. The second model predicts that UGGT recognizes the innermost *N*-acetylglucosamine (GlcNac) molecule of the core glycan. In folded proteins, this amino sugar interacts with amino acid residues lying in its vicinity. This interaction may be impaired upon protein misfolding and may consequently lead to UGGT recognition (Parodi 2000). A third model proposes that UGGT may recognize specific dynamic properties exhibited only by non-native glycoprotein conformations.

In addition to CNX and CRT, several other chaperones and folding factors, such as the BiP protein, are involved in the ER quality control of membrane proteins. The use of the different chaperones by the individual proteins is very variable. Some membrane proteins, such as the vesicular stomatitis virus (VSV) protein G (Hammond and Helenius 1994a) or the hepatitis B virus protein M (Mehta et al. 1997), require the CNX/CRT system for correct folding. The G protein-coupled V_2R, in contrast, does not require the lectine chaperones for correct folding (Innamorati et al. 1996). Nevertheless, lectine chaperones play a role in the ER retention of misfolded receptor forms (Morello et al. 2001). For other glycoproteins, it was demonstrated that, if one chaperone is inactive, another one can take over its function. The influenza hemagglutinin (HA) protein, for example, was found to associate with BiP, if the CNX/CRX interaction is inhibited (Zhang et al. 1997).

How do membrane proteins choose among the different chaperones? Molinari and Helenius (2000) proposed a simple mechanism for the choice of glycoproteins between CNX/CRT and BiP. They demonstrated that proteins interact with the CNX/CRT system only if the N-glycosylation consensus site is contained within the first 50 amino acid residues of the N terminus. If the glycosylation site is located more C-terminally, proteins interact with BiP.

Signal-based quality control

The chaperone-mediated recognition of misfolded proteins through exposed hydrophobic domains or via the innermost GlcNac unit is relatively nonspecific, enabling chaperones to interact with a broad range of target proteins. An additional, more specific concept was proposed by Zerangue et al. (1999). During a study of the assembly and trafficking of ATP-sensitive potassium channels, an ER retention signal with the sequence RXR was identified in both the α and β subunits of the octameric channel complex. When the individual subunits were expressed alone, they were retained in the ER via these RXR sequences. Coexpression of the subunits and channel oligomerization, however, allowed their transport to the cell surface. It was proposed that the RXR signals are exposed and recognized by adapter proteins in the ER as long as the channel subunits are not properly assembled. Upon complete channel assembly, however, they may be shielded. Such a mechanism would ensure that only fully assembled channel complexes reach the plasma membrane. RXR signals may thus contribute to a novel signal-based quality control mechanism which is necessary to monitor the assembly of multimeric proteins.

RXR signals with a similar function were recently described in the subunits of the *N*-methyl-*d*-aspartate (NMDA) receptor channel complex (Scott et al. 2001). RXR signals may also regulate the transport of dimeric GPCRs. The gamma amino butyric acid receptor subtype B ($GABA_B$ receptor) occurs in two different subtypes, $GABA_{B1}$ and $GABA_{B2}$. The $GABA_{B1}$ receptor is retained in the ER via an RXR retention signal located in its C-terminal tail when expressed alone. Heterodimerization with the $GABA_{B2}$ receptor, however, masks the RXR signal and allows the transport of the heterodimer to the plasma membrane (Margeta-Mitrovic et al. 2000; Pagano et al. 2001).

The results discussed so far show that RXR motifs play a role in monitoring the assembly of multimeric proteins. The postulated mechanism of shielding these motifs only upon complete assembly provides a powerful tool to avoid the transport of unassembled protein complexes. Recent results suggest that RXR motifs may also play a role in the quality control of monomeric folding-defective proteins. The ΔF508 mutant of the CFTR protein, which causes cystic fibrosis in the majority of patients, is retained in the early secretory pathway. Mutation of all putative RXR sequences on the cytoplasmic side leads to a partial rescue of the cell surface transport of this protein. These results are consistent with a function of RXR motifs in the quality control of mutant CFTR proteins (Chang et al. 1999). A similar role of RXR motifs was suggested for the G protein-coupled V_2R (Hermosilla and Schülein 2001).

Taken together, the available data suggest that signal-mediated pathways play an important role in the quality control of both protein oligomerization and folding. The proteins recognizing the exposed RXR motifs in the ER are not very well characterized. Recent data indicate that the coat protein complex I (COP I; see below) and the 14–3-3ζ protein may be involved (Yuan et al. 2003).

Quality control in the ERGIC

Correctly-folded proteins are allowed to enter the vesicular transport from the ER to the Golgi apparatus via the ERGIC (Klumperman 2000; Hauri et al. 2000; Antonny and Schekman 2001; Fig. 4). They are packed into the membrane of budding vesicles (Schekman and Orci 1996) leaving the ER at ribosome-free transitional elements.

Vesicles leaving the ER contain a coat protein complex II (COPII) coat. Budding is driven by COPII polymerization, i.e., by the sequential interaction of the individual COPII components on the surface of the vesicles (Sarl GTPase, Sec23p/Sec24p and Sec13p/Sec31p heterodimeric complexes) (Springer et al. 1999). Cargo loading into the COPII vesicles is either signal-mediated (Nishimura and Balch 1997, Kuehn et al. 1998; Hermann et al. 1999; see also "Leaving the ER: specific recruitment vs. bulk flow") or occurs by bulk flow (Wieland et al. 1987; Martinez-Menargez et al. 1999). Soon after budding, the vesicles lose their coat and fuse with the ERGIC, the first sorting station for proteins on their way to the Golgi apparatus. The ERGIC (also known as VTCs=vesicular tubular clusters) is composed of highly mobile tubovesicular structures located in the vicinity of the Golgi apparatus and in the cell periphery. The ERGIC may be formed by fusion of the arriving transport vesicles (maturation hypothesis) or may pre-exist as a stable compartment (stable compartment hypothesis). The ERGIC is not only a sorting station for secretory and membrane proteins on their way to the Golgi apparatus (anterograde transport), but also for proteins transported back to the ER (retrograde transport; Aridor et al. 1995; Klumperman et al. 1998; Hauri et al. 2000). In the case of the retrograde transport from the ERGIC to the ER, proteins are packed into coat protein complex I (COPI)-coated vesicles. The COPI coat consists of the small GTPase ADP-ribosylation factor (ARF) and a preassembled coatomer complex of seven subunits (Rothman 1994).

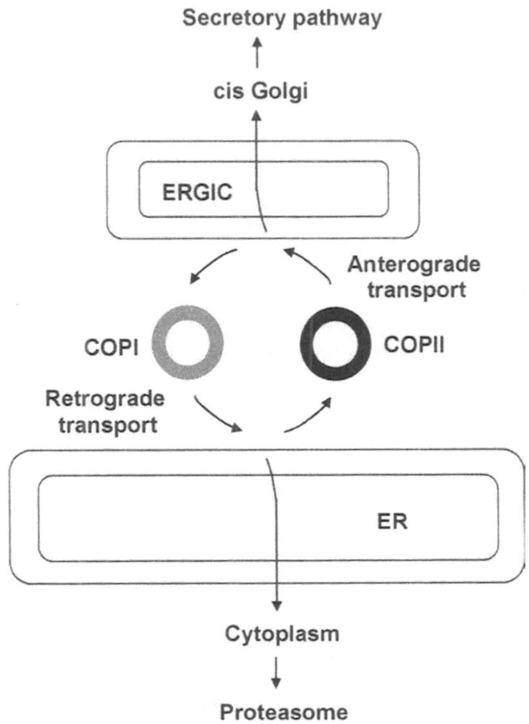

Fig. 4 Anterograde and retrograde transport between the ER and the Golgi apparatus. See text for details

It was postulated that the normal function of the ERGIC is to concentrate proteins in the secretory pathway. This may be achieved by excluding them from the retrograde transport (see "Recruitment mechanisms of soluble proteins"). A growing number of studies indicates that the ERGIC also plays a role in the quality control of misfolded proteins. Hammond and Helenius (1994b) demonstrated that a temperature-sensitive mutant of the VSV-G protein is able to leave the ER and to reach the ERGIC. The VSV-G protein is then routed back to the ER via the retrograde transport system. These results demonstrate that the ERGIC is able to recognize at least some misfolded proteins. A cycling between the ER and ERGIC was also observed for the cystic fibrosis-causing ΔF508 mutant of the CFTR protein (Gilbert et al. 1998). The ERGIC appears to recognize not only misfolded proteins, but also unassembled subunits of multimeric proteins, e.g., subunits of the T-cell antigen receptor (TCR; Yamamoto et al. 2001).

How are misfolded proteins recognized in the ERGIC, and how are they sorted into the COPI-coated retrograde transport vesicles? In the case of the unassembled TCR α chain, it was postulated that the retrograde transport is mediated by association of the α chain with BiP (Yamamoto et al. 2001). BiP contains a C-terminal KDEL retrieval signal that enables the BiP/α chain complex to enter the retrograde COPI-coated vesicles by binding to the KDEL receptor (Yamamoto et al. 2001).

Whereas the data discussed above strongly suggest that the ERGIC is part of the quality control system, its precise function within this system is not understood. It is not clear, for example, why some misfolded proteins reach the ERGIC, whereas others remain in the ER. In addition, it is an open question whether different misfolded mutants of the same protein behave identically, i.e., whether they all reach the ERGIC or whether some mutants reach the ERGIC and others remain in the ER. Molinari and Helenius (2000) proposed that an individual membrane protein has only a very limited ability to choose between different quality control mechanisms. For example, the interaction with the CNX/CRT system or BiP is determined by the location of the glycosylation site within the N tail (see "Chaperone-mediated quality control: the CNX/CRT cycle"). This suggests that different mutants of the same glycoprotein are not recognized by different components of the quality control system. However, for the V_2R, it was recently shown that the L62P mutant is retained exclusively in the ER, whereas the Y205C mutant reaches the ERGIC demonstrating that different mutants of the same protein are retained intracellularly by different quality control mechanisms (Hermosilla et al., manuscript in preparation). The ability of a particular protein to choose between different quality control mechanisms thus appears to be less limited than previously thought.

Leaving the ER: specific recruitment vs. bulk flow

The question of whether the transport out of the ER occurs by bulk flow or by specific recruitment leading to cargo concentration in the vesicles has been a matter of controversy. Recently, it became clear that both mechanisms exist: many if not all membrane proteins are actively recruited into the vesicles by direct interactions with the COPII coat (Fig. 5, left). Some soluble proteins may also be actively sorted, but only with the help of specific cargo receptors (Fig. 5, middle). The majority of the soluble proteins, however, appear to leave the ER by bulk flow (Fig. 5, right).

Recruitment mechanisms of membrane proteins

Most membrane proteins are expressed at a very low level. It was thus reasonable to postulate specific, signal-based recruitment mechanisms at the ER exit sites, leading to vesicular concentration. In principle, membrane proteins may interact directly or indirectly with the COPII coat. In the case of direct interactions, the sorting signals are contained in

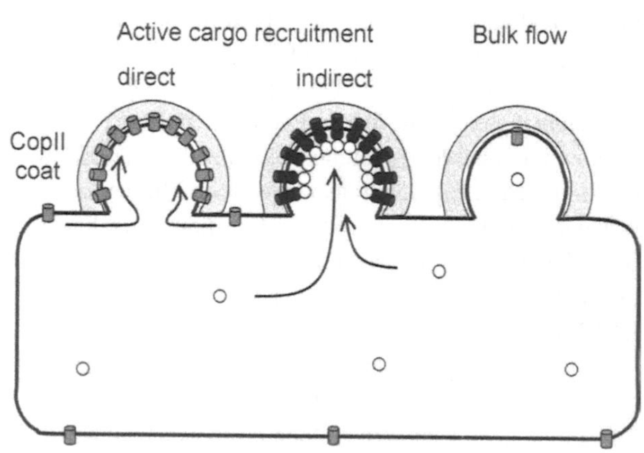

Fig. 5 Recruitment of cargo proteins into COPII-coated vesicles. *Left*: Active cargo recruitment. Membrane proteins (*dark gray*) bind directly to the COPII coat (*light gray*). Soluble cargo proteins (*white*) may be sorted indirectly by binding to membrane-bound cargo receptors (*black*). Both direct and indirect sorting mechanisms lead to cargo concentration. *Right*: Soluble proteins may also enter the vesicles by bulk flow rather than by sorting. In this case, they are not further concentrated. See text for details

the cargo proteins themselves. In the case of indirect interactions, the sorting signals are contained in cargo receptors which bind to both the COPII coat and the cargo proteins. To date, only direct sorting has been demonstrated for membrane proteins. This does not preclude that at least some membrane proteins are sorted by cargo receptors as has been shown for soluble proteins (see below).

Direct binding of a membrane protein to the COPII coat had been demonstrated, e.g., for the yeast *N*-ethylmaleimide-sensitive factor attachment protein (SNARE) Bet1. This protein specifically interacts via its N-terminal cytosolic domain with the Sec24p subunit of the Sec23p/Sec24p heterodimeric complex (Springer and Schekman 1998, Miller et al. 2003). Other SNAREs, like Sec22p and Ykt6p, may interact more weakly with the COPII coat (Springer and Schekman 1998). The function of these SNARE proteins is to confer fusion capacity to vesicles, i.e., they play a role in ER to Golgi traffic. These proteins are thus "frequent flyers" of the vesicles, and it is believed that their COPII-interacting domains confer sorting privileges (Antonny and Schekman 2001). Another protein belonging to this group is the yeast *cis*-Golgi syntaxin Sed5p, which specifically interacts with Sec24p (Peng et al. 1999).

The COPII-interacting sequence motifs of membrane proteins were characterized in the case of the lectine ERGIC-53 (see "Quality control in the ERGIC"), a cargo receptor for glycoproteins. ERGIC-53 is also one of the frequent flyer proteins of the anterograde transport vesicles requiring sorting privileges (see "Recruitment mechanisms of soluble proteins"). Binding of ERGIC-53 to the COPII coat is mediated by a C-terminal diphenylalanine ER exit signal (Kappeler et al. 1997; Nufer et al. 2003). The phenylalanine residues of the motif in the human protein are replaced by leucine residues in yeast, suggesting that the hydrophobicity of the motif rather than its specific amino acid sequence is required for COPII binding (Hauri et al. 2000). Diphenylalanine motifs also stimulate the ER export of proteins belonging to the p24 family of putative cargo receptors (Fiedler et al. 1996).

Recent data indicate that ER exit signals are not restricted to proteins of the early secretory pathway requiring sorting privileges. The VSV-G protein, for example, is very efficiently concentrated in COPII-coated vesicles by a diacidic C-terminal EXD signal binding to Sar1p and the Sec23p/Sec24p heterodimer (Nishimura and Balch 1997; Nishimura et al. 1999; Aridor et al. 2001). This signal is also transferable to other proteins (Sevier et al. 2000). Diacidic motifs appear to be contained in many other unrelated membrane proteins (Bannykh et al. 1998). In yeast, a DXE motif stimulates the ER export of the Sys1p protein by binding to the Sec23p/Sec24p heterodimer (Votsmeier and Gallwitz 2001). In the C tail of inwardly rectifying potassium channels, diacidic motifs with the sequences EXD and EXE were identified which have a similar function (Ma et al. 2001; Stockklausner et al. 2001). Recently, a dibasic motif (RKXRK) was shown to mediate ER export of the Golgi resident glucosyltransferases by binding to the Sar1 GTPase of the COPII coat (Giraudo and Maccioni 2003).

The data discussed above indicate that not only the proteins of the early secretory pathway but also many other membrane proteins enter the vesicular transport by specific recruitment mechanisms rather than by bulk flow.

Recruitment mechanisms of soluble proteins

Some secreted proteins are very efficiently sorted into COPII vesicles. It has been suggested that they bind indirectly to the COPII coat via cargo receptors containing COPII-interacting domains. The protein ERGIC-53 belongs to the group of these putative cargo receptors. Several properties of ERGIC-53 suggest such a function. Firstly, the protein cycles between the ER and the ERGIC (Aridor et al. 1995; Klumperman et al. 1998). It is thus not lost after ER exit but can be retrieved. Secondly, the protein contains a large luminal domain that may interact with soluble cargo proteins. The domain is homologous to the carbohydrate recognition domain of plant lectines and the human vesicular-integral membrane protein (VIP) 36 (Fiedler et al. 1994). Recent studies strongly suggest that ERGIC-53 functions as a cargo receptor for at least some glycoproteins. Using chemical crosslinking, a glycoprotein related to human cathepsin Z was identified as a ligand for ERGIC-53 (Appenzeller et al. 1999). In addition, mutations in ERGIC-53 impaired the transport of the human coagulation factors V and VIII, thereby causing the autosomal recessive bleeding disorder (combined deficiency of coagulation factors V and VIII; Nichols et al. 1998; Nichols et al. 1999; Neermann-Arbez et al. 1999; Moussalli et al. 1999). These data suggest that the two coagulation factors need ERGIC-53 as a cargo receptor for their transport through the early secretory pathway.

The Emp24p protein of yeast represents a putative cargo receptor that does not belong to the lectine family. Its knockout delays the transport of invertase and the glycosylphosphatidylinositol (GPI)-anchored Gas1 protein (Muniz et al. 2000). The yeast protein Erv29p is another cargo receptor; it enriches the glycosylated pro-alpha-factor (gpalphaf) approximately 20-fold in COPII vesicles (Belden and Barlowe 2001; Malkus et al. 2002).

The available data suggest that soluble proteins may also be sorted actively at the ER exit sites. These results are inconsistent with a generalized bulk flow concept, postulating that they enter COPII vesicles in a nonspecific manner and without further concentration (Wieland et al. 1987). However, recent work suggests that such a mechanism also exists. Martinez-Ménarguez et al. (1999) demonstrated that two secretory proteins of exocrine pancreatic cells, amylase and chymotrypsinogen, leave the ER by bulk flow. Concentration measurements by quantitative immunelectron microscopy revealed equal concentrations of both proteins in the ER lumen, the transition elements, COPII-coated buds, and vesicles. Both proteins were, however, strongly concentrated in the ERGIC, most likely because they are excluded from the retrograde transport into COPI vesicles. In the same study, concentration measurements were also performed with two membrane proteins as controls, namely the KDEL receptor and the vSNARE protein Bet1. In contrast to the secretory proteins, the concentration of the membrane proteins showed a significant increase in COPII-coated ER buds and in anterograde vesicles.

In summary, the available data suggest that at least some soluble proteins are actively recruited into COPII-coated vesicles, whereas others enter them by bulk flow. Further studies are needed to characterize the properties of the proteins requiring one or the other pathway. Interestingly, bulk flow was observed for two abundant proteins, amylase and chymotrypsinogen (Martinez-Ménarguez et al. 1999). It may thus be speculated that the expression level of a soluble protein determines the ER exit mechanism. It is conceivable that active recruitment is advantageous for proteins expressed at low levels since such ā mechanism would allow their concentration at the ER exit sites.

Transport signal or folding-relevant domain: the difficulties of defining ER exit signals in membrane proteins

ER exit signals of membrane proteins mediate sorting into the anterograde transport vesicles by binding to COPII. Experimentally, ER exit signals are identified by site-directed mutagenesis and subsequent transport studies. The intracellular location of the mutants is assessed by glycosylation state analyses and/or immunofluorescence microscopy. The identification of novel ER exit signals of membrane proteins is complicated by the fact that mutations impairing protein folding may also cause transport defects. ER exit signals must thus be differentiated from domains that are relevant for transport-competent folding. The following section will discuss experimental approaches to identify ER exit signals. What are the basic characteristics of ER exit signals, and how can they be distinguished from folding-relevant domains?

a. ER exit signals work independently of the rest of the protein, and their transport functions are transferable to other proteins. In contrast, folding-relevant domains are not transferable and function only if they are embedded in their original sequence background.
b. The currently known ER exit signals are recognized by COPII components (see above). As the COPII coat is formed on the vesicle surface, ER exit signals must be located on the cytoplasmic side of a membrane protein. In contrast, folding-relevant domains contribute to intramolecular interactions and are not restricted to the cytoplasmic face of membrane proteins.
c. Mutation of ER exit signals leads to functional proteins retained in the ER. In contrast, mutations in folding-relevant domains normally lead to transport-defective, nonfunctional forms.
d. The most striking feature of the currently known ER exit signals is their size. The signals (FF, LL, DXE, see above) are rather small and there is only limited support for the idea that residues in the vicinity of the signals may contribute to their function (e.g., Ma et al. 2001). The size of the ER exit signals is reminiscent of descriptions of other sorting signals involved in later transport steps. Among these are the basolateral sorting signals of the tyrosine type (consensus sequence NPXXY or YXXØ), first described for the low-density lipoprotein (LDL) receptor (Matter et al. 1992), or of the dileucine type (consensus LL), first described for the immunoglobulin G (IgG) Fc receptor (Hunziker and Fumey 1994). These two signals are recognized in the *trans*-Golgi network by the clathrin-associated adaptor protein complex 1 (AP1) (Yeaman et al. 1999). The tyrosine motif is bound by the μ chain, and the dileucine motif by the *β* chain of AP1. The crystal structure of the μ chain in complex with a tyrosine motif revealed that the sorting signal is bound in a non-folded, extended conformation, which explains how it can function as such a small sequence (Owens and Evans 1998). If one assumes that the ER exit signals are also recognized in an extended conformation, their small dimensions become no longer surprising. Sequences relevant for transport-competent folding should also be distinguishable by their size. They should be significantly larger than ER exit signals since they are embedded into structured domains.

The difficulties of distinguishing ER exit signals from folding-relevant domains became evident in studies addressing the transport relevance of the C tails of GPCRs. Several C-terminally truncated GPCRs are retained in the ER. Among these are the rat luteinizing

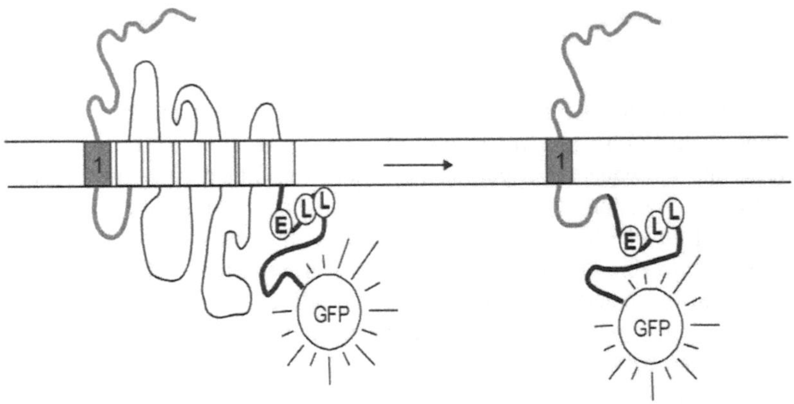

Fig. 6 Development of an experimental system to study the transport functions of the glutamate/dileucine motif of the V_2R independently of full-length receptor folding. The C terminus of the receptor (*black*) was fused to a fragment containing only the N tail, TM1, and ICL1 (*gray*). Mutations of the glutamate/dileucine motif did not influence the transport of the construct shown on the right side

hormone/choriogonadotropin (LH/CG) receptor (Rodriguez et al. 1992), the rat glucagon receptor (Unson et al. 1995), and bovine rhodopsin (Heymann and Subramaniam 1997).

In the case of the V_2R, a sequence in the C tail required for ER exit was characterized in detail. The receptor contains two palmitoylated cysteine residues in its C tail (C^{341} and C^{342}). Mutation of these cysteines reduced the receptor transport to the plasma membrane (Schülein et al. 1996b). Deleting only four additional residues N-terminally of the cysteines by truncation at R^{337} abolished receptor transport to the plasma membrane completely (Sadeghi et al. 1997; Oksche et al. 1998). These results strongly suggested that sequences N-terminally of the palmitoylated cysteine residues play a crucial role for the ER exit of the receptor. Indeed, it was shown that a glutamate/dileucine motif (^{335}<u>E</u>LRS<u>LL</u>CC^{343}) lying immediately N-terminally of the palmitoylated cysteines is necessary for the ER exit of the V_2R (Schülein et al. 1998a).

At first glance, the glutamate/dileucine motif of the V_2R fulfilled all the requirements for a ER exit signal: it was small, located in a cytoplasmic domain, and it was similar to the dihydrophobic ER exit signals described so far. However, when the C tail of the V_2R was fused to a V_2R fragment comprising only the N tail, first transmembrane domain and first cytoplasmic loop, i.e., when the C tail was expressed independently of full-length receptor folding, mutations in the glutamate/dileucine motif did not cause transport-defects (Krause et al. 2000; Fig. 6). These results show that the glutamate/dileucine motif of the V_2R represents a domain relevant for transport-competent receptor folding rather than an ER exit signal. This was surprising, since it was previously thought that folding of GPCRs and other membrane proteins is mainly driven by the interactions of the TMs and that the cytoplasmic domains have only a minor influence.

Recently, it was postulated that a conserved hydrophobic sequence in the C tail of the D_1 dopamine receptor may nonetheless represent an ER exit signal (Bermak et al. 2001). In contrast to the V_2R, the sequence was not only transport-relevant for the full-length receptor, but also when fused to a single spanning receptor fragment (see Fig. 6 for the original V_2R experiment). Mutation of the hydrophobic domain led to an ER-retained receptor fragment that was recognized by a new ER-resident protein, DRiP78, whose function is currently unknown (Bermak et al. 2001). However, the approach to express and to mutate a sequence independent of full-length receptor folding leads to unambiguous results only if the transport of the mutant receptor fragments is preserved, i.e., if the domains in question do not contain transport signals. Conversely, an impaired transport does not verify the existence of transport signals, because it cannot be ruled out that the transport defects result again from misfolding, not of the full-length receptor in this case, but of the receptor fragments. Thus, an alternative interpretation of the data obtained for the D_1 dopamine receptor is that mutation of the hydrophobic sequence in the receptor fragment leads to a misfolded fragment that is recognized by the quality control system. If so, the DRiP78 protein may represent a new chaperone that retains the misfolded construct.

The examples of the V_2R and the D_1 dopamine receptor demonstrate the difficulties in identifying new ER exit signals in membrane proteins. For the future, it is necessary to include structural data in the search for ER exit signals. Such information is crucial to judge whether or not the domain in question is folding-relevant.

Examples of diseases caused by the retention of membrane proteins in the early secretory pathway

The study of the early secretory pathway of membrane proteins has major clinical implications. Mutations in the genes of membrane proteins frequently lead to misfolded proteins that are recognized by the quality control system and retained intracellularly. Such mutations are the molecular basis of many inherited diseases ("conformational diseases").

The list of such diseases has grown considerably during the past years. This section will focus on diseases for which the trafficking defects and/or the components of the quality control system involved have been studied in more detail (see the sections entitled "Cystic fibrosis," "Congenital nephrogenic diabetes insipidus," "Retinitis pigmentosa," "Familial hypercholesterolemia," "Congenital sucrase-isomaltase deficiency," "Hereditary hemochromatosis," "Oculocutaneous albinism," "Laron syndrome"). A second focus is on diseases where the trafficking studies led to a better understanding of the mechanisms of dominant-negative inheritance (see "Congenital nephrogenic diabetes insipidus," "Charcot-Marie-Tooth syndrome," "Congenital long QT syndrome").

A comprehensive list of diseases associated with protein transport defects (including secretory proteins) was presented by Aridor and Hannan (2000). The updated database is also available online at www.traffick.dk (click on the toolbox icon). To date, transport studies are only available for a few of the diseases known to be caused by mutations in the genes of membrane proteins. Thus, it is conceivable that the list of diseases associated with transport-deficient membrane proteins will grow rapidly in the near future. Disorders belonging to this group may include in particular the various inherited "channelopathies" (Bockenhauer 2001; Celesia 2001; Hübner and Jentsch 2002).

Cystic fibrosis

Cystic fibrosis (CF) is the most common fatal autosomal recessive genetic disorder (60,000 patients worldwide; incidence in the US: 1:1.3,500 live Caucasian births, 1:25,000 live Asian births). The disease manifests in its classic and most common form as a chronic obstructive lung disease accompanied by pancreatic insufficiency. Patients suffer in particular from an increased production of viscous mucus favoring the occurrence of severe respiratory infections.

Cystic fibrosis is caused by mutations in the gene for the CFTR protein, a cAMP-regulated chloride channel belonging to the ABC transporter subfamily (Sheppard and Welch 1999; Schwiebert et al. 1999; Zielenski 2000; Ko and Pedersen 2001). CFTR is a monomeric type II membrane protein consisting of 1480 amino acids. It contains two large membrane-spanning complexes (each consisting of 6 TMs), two nucleotide-binding domains, and a central, highly charged regulatory domain with multiple phosphorylation consensus sites.

The CFTR protein is expressed in subapical, endosomal compartments and in the apical membrane of secretory epithelial cells (Bradbury 1999). The principal function of the CFTR protein is in its role in the cAMP-regulated chloride transport across the apical membrane of epithelial cells, but it has also been implicated in the regulation of other ion channels, membrane trafficking, pH regulation, and apoptosis.

To date, 1,291 different mutant CFTR alleles have been reported (cystic fibrosis genetic analysis consortium, 11/2003; website: www.genet.sickkids.on.ca/cftr/). The mutations are classified into 6 subgroups:

Class 1 mutations abolish either the synthesis of the CFTR protein or lead to a completely altered protein structure. This group encompasses the nonsense, frameshift, RNA splicing, and the missense mutations changing the translation initiation codon.

Class 2 mutations lead to abnormal processing and trafficking of the CFTR protein. These mutant proteins are normally retained in the early secretory pathway.

Class 3 mutations affect the regulation of the CFTR protein. Mutant proteins are, e.g., unable to bind and to hydrolyze ATP.

Class 4 mutations cause a decrease in channel conductance and, because of the residual function, are generally associated with a milder phenotype.

Class 5 mutations lead to a fully functional CFTR protein. The total amount of the protein, however, is reduced. The promoter mutations upstream of the CFTR gene, which are normally associated with a milder phenotype, belong to this group.

Class 6 mutations represent a novel subgroup. Here, stop codons lead to truncated C tails in the CFTR protein. Mutant proteins are fully functional but very unstable (Haardt et al. 1999).

Class 2 mutations are the most frequent in cystic fibrosis. This group includes the ΔF508 mutation (Riordan et al. 1989), which is found in about 70% of the patients. The ΔF508 mutation is a three-base pair deletion in the sequence encoding the first nucleotide binding site. Due to its significance for cystic fibrosis, the functional consequences of the ΔF508 mutation have been studied intensively. The mutant protein is incompletely processed, retained in the ER (Cheng et al. 1990; Denning et al. 1992) and finally subjected to proteolysis (Jensen et al. 1995). These data indicate that the ΔF508 mutation causes a folding defect recognized by the quality control system. Several proteins were described as being involved in this process. The chaperones calnexin (Pind et al. 1994), Hsp70 (Yang et al. 1993) and the Hsc70/Hdj-2 pair (Meacham et al. 1999) display a prolonged association with the mutant protein. Furthermore, ER retention signals of the RXR sequence-type appear to be exposed upon misfolding and may play a role in the retention process (Chang et al. 1999). Recent data show that the quality control of the ΔF508 mutant is not restricted to the ER. A substantial portion of the protein reaches the ERGIC and is then directed back to the ER via the retrograde transport pathway (Gilbert et al. 1998; Bannykh et al. 2000).

In summary, quality control of the ΔF508 mutant involves chaperones, signal-based mechanisms, and the ERGIC—i.e., all three mechanisms currently known. Most interestingly, the ΔF508 mutant is functional in the ER membrane (Pasyk and Foskett 1995) although the protein is less stable than the wild type channel (Sharma et al. 2001; Heda et al. 2001). Quality control thus appears to be overprotective in this case.

The ΔF508 CFTR mutant has become a model protein for studying the intracellular retention of membrane proteins as a disease-causing mechanism. Furthermore, the majority of attempts to correct misfolding and mislocalization of membrane proteins by pharmacological means have been made with this mutant (see "Pharmacological approaches to correct membrane protein misfolding and retention in the early secretory pathway").

Congenital nephrogenic diabetes insipidus

Congenital nephrogenic diabetes insipidus (NDI) is a rare disorder (incidence 1:100.000–1:500.000) characterized by the kidney's inability to respond to the antidiuretic hormone 8-arginine vasopressin (AVP; Oksche and Rosenthal 1998; Knoers and Deen 2001; Morello and Bichet 2001). The consequence is the production of a large amount of dilute urine. Due to the massive loss of water, the patients suffer from thirst and are in danger of dehydration. The disease usually becomes evident shortly after birth. In addition to the primary NDI symptoms of polyuria and polydipsia, symptoms and signs such as fever, irritability, constipation, failure to thrive, lack of appetite, vomiting, and high blood levels of sodium may be observed.

NDI-causing mutations have been found in the genes encoding the V_2R (Rosenthal et al. 1992) and the water channel protein aquaporin (AQP) 2 (Deen et al. 1994). Both proteins are expressed in the principal epithelial cells of the renal collecting duct and are key components of AVP-mediated signal transduction (Fig. 7).

The V_2R is mainly expressed in the basolateral membrane. Upon AVP binding, the receptor couples to the G_s/adenylyl cyclase system. The subsequent rise in intracellular cAMP levels leads to the activation of the protein kinase A (PKA) and to the fusion of AQP2-containing vesicles with the apical membrane by an as yet incompletely understood mechanism (AVP-mediated exocytosis). The presence of AQP2 in the apical membrane increases the luminal water permeability of the epithelium. The water is absorbed into the hypertonic interstitium through AQP3 and AQP4, which are constitutively expressed in the basolateral membrane.

At first glance, the AVP-mediated exocytosis is reminiscent of the regulated exocytosis in neurones. Here, the fusion of transmitter-loaded vesicles with the synaptic membrane leads to transmitter secretion. However, whereas the neuronal exocytosis is calcium-dependent and occurs within milliseconds, the AVP-mediated exocytosis appears to be independent of intracellular calcium and a slow process completed within minutes (Lorenz et al. 2003)

NDI-causing V2R mutations. The V_2R consists of 371 residues and belongs to the large subfamily of rhodopsin-like GPCRs (family 1 according to the classification of Bockaert and Pin 1999). More than 150 different NDI-causing V_2R mutations have been described.

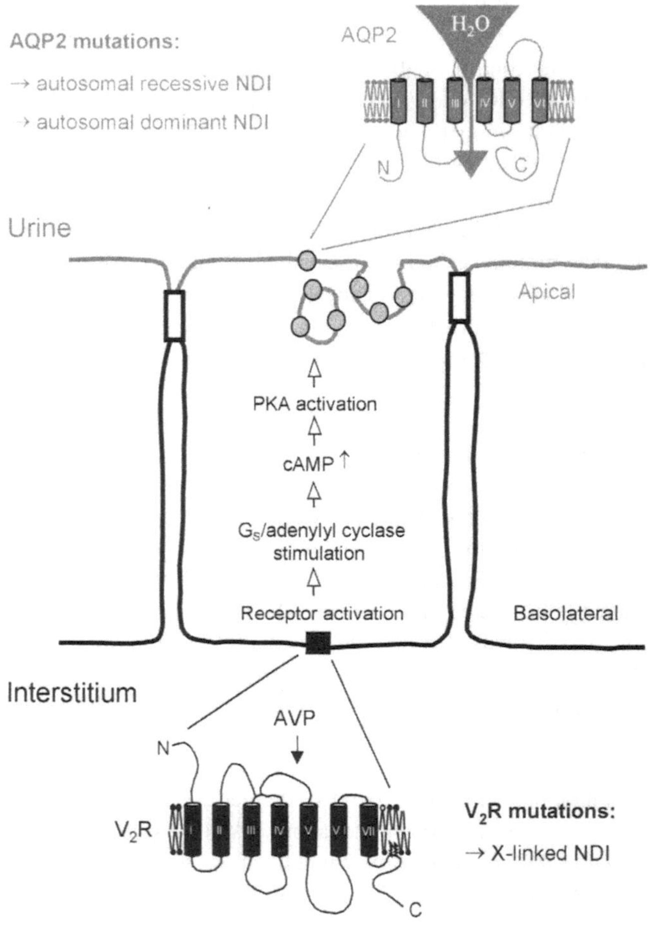

Fig. 7 AVP-mediated signal transduction in the principal epithelial cells of the renal collecting duct. The signal transduction is polarized and proceeds from the basolateral (*black*) to the apical (*gray*) side of the epithelium. AVP stimulation of the V_2R in the basolateral membrane leads to the insertion of AQP2-containing vesicles into the apical membrane, thereby increasing water permeability (see text for details). NDI-causing mutations were found in genes encoding the AQP2 protein (autosomal dominant and recessive forms) and the V_2R (X-linked form). The topology of both proteins is indicated. The AQP2 protein is a hexahelical type II membrane protein. The V_2R is a heptahelical type III membrane protein

They are responsible for about 90% of all congenital NDI cases (Oksche and Rosenthal 1998; Morello and Bichet 2001).

In the case of the V_2R, NDI is transmitted as an X-linked recessive trait. Many NDI-causing mutant V_2Rs were analyzed functionally. Currently, three classes of V_2R mutants are distinguished: class 1 mutants are defective in ligand-binding or G protein-coupling, class 2 mutants are characterized by an impaired intracellular transport, and class 3 mutants display decreased biosynthesis levels.

The majority of the V_2R mutations lead to transport-deficient receptors (e.g., Oksche et al. 1996; Morello and Bichet 2001). These mutations are typically located in the transmembrane domains. Many of them introduce or eliminate proline residues (e.g., the mutants L59P, L62P, L135P, R143P, L219P, L282P, L292P, A294P and P95L, P173S, P217T, P322S). It is conceivable that these mutations impair helical structures and consequently cause folding defects recognized by the quality control system. During the past decade, the NDI-causing V_2R mutants became an important model for the study of the quality control mechanisms of membrane proteins. Similarly to the situation with the CFTR protein (see "Cystic fibrosis"), at least three different mechanisms contribute to the V_2R quality control:

a. Recognition by the CNX/CRT system. For the truncation mutant R337X, a prolonged CNX association was observed, indicating that lectine chaperones play a role in the retention of misfolded V_2R mutants (Morello et al. 2001). Interestingly, elimination of the single glycosylation site of the V_2R (N^{22}) had no influence upon receptor function, demonstrating that the CNX/CRT system is not required for correct receptor folding (Innamorati et al. 1996).
b. Recognition by exposed ER retention signals. The V_2R contains a putative RXR-type retention signal in its ICL3 domain. It was proposed that this signal is exposed in some misfolded V_2R mutants and subsequently recognized by the quality control system (Hermosilla and Schülein 2001; see "Signal-based quality control").
c. Recognition in the ERGIC. The ERGIC recognizes misfolded receptors that have escaped from the ER by an as yet unknown mechanism. The mutant receptors are then directed back to the ER via the retrograde transport pathway (Hermosilla et al., manuscript in preparation; see "Quality control in the ERGIC").

Folding-relevant regions are not only located in the TMs of the V_2R. In the soluble domains, the putative α helix in the C tail (see "Transport signal or folding-relevant domain: the difficulties of defining ER exit signals in membrane proteins") and the conserved disulfide bond between C^{112} and C^{192} were also shown to be necessary for a transport-competent folding state. Mutation of either one of the two cysteines constituting the disulfide bond led to intracellularly retained receptors. It was speculated that NDI-causing mutants with additional extracellular cysteine residues (R106C, R181C, G185C, R202C, Y205C) form an alternative disulfide bond with either C^{112} or C^{192} (van den Ouweland et al. 1992; Pan et al. 1994; Yokoyama et al. 1996; Ala et al. 1998). If this hypothesis holds true, the "additional C" mutations should also lead to class 2 transport-defective receptors. This was, however, not the case. Although binding-defective, the mutants G185C and R202C were efficiently targeted to the plasma membrane (class 1 mutants; Schülein et al. 2000). Recently, it was shown that the additional cysteine residues of these mutants form a second disulfide bond with a nonconserved extracellular cysteine, rather than impair the conserved bond (Schülein et al. 2001).

All transport-defective V_2R mutants described so far were retained in the early secretory pathway. This is surprising, since the later vesicular transport steps of the receptor are rather complex. For example, the receptor is enriched in the basolateral subdomain of the epithelial cells (Schülein et al. 1998b; Andersen-Beckh et al. 1999; Hermosilla and Schülein 2001). It is conceivable that this polarized expression requires additional sorting information which may be lost as a consequence of NDI-causing mutations. However, such mutations have not been identified yet.

NDI-causing AQP2 mutations. The AQP2 protein belongs to the family of aquaporin water channels (Fushimi et al. 1993). It is a hexahelical type II membrane protein consisting of 271 amino acids. Aquaporins form homotetramers, each monomer containing a distinct water pore (Verkman and Mitra 2000; Nielsen et al. 2002). Thirty-five different NDI-causing AQP2 mutations have been described. They are responsible for about 10% of all congenital NDI cases (Knoers and Deen 2001). Twenty-seven mutations cause an autosomal recessive, 8 mutations an autosomal dominant form of the disease.

The group of AQP2 mutations causing autosomal recessive NDI include the mutants G64R, R187C, S216P (Deen et al. 1995), A147T, T126M, N68S (Mulders et al. 1997), C181W (Canfield et al. 1997), and L28P, A47V, V71M, T125M, G175R P185A (Marr et al. 2002a). They are retained in the ER by the quality control system although they may have conserved some functional activity (Tamarappoo and Verkman 1998; Tamarappoo et al. 1999; Marr et al. 2002a). The R187C mutant is retained in a monomeric form and is unable to form tetramers with wild-type AQP2 (Kamsteeg et al. 1999). These results explain the recessive inheritance of this mutant (Kamsteeg et al. 1999).

The AQP mutations causing autosomal dominant NDI are found in the sequence encoding the C tail of the protein. Examples are the mutants E258K (Mulders et al. 1998; Kamsteeg et al. 1999) and three C-terminal deletion mutants (Kuwahara et al. 2001). In contrast to the AQP2 mutants causing recessive NDI, the E258K mutant was not retained in the ER but in the Golgi apparatus (Mulders et al. 1998, Kamsteeg et al. 1999). It formed tetramers with wild-type AQP2 thus explaining its dominant inheritance (Kamsteeg et al. 1999).

The results obtained for the NDI-causing AQP2 mutants not only indicate that AQP2 tetramerization takes place in the Golgi apparatus, but also that this compartment is involved in the quality control of AQP2 channel assembly. Thus, AQP2 is the one of the few membrane proteins whose quality control is not restricted to the early secretory pathway (see also the data for the sucrase isomaltase enzyme in the section "Sucrase-isomaltase deficiency"). Recently, examples of autosomal-dominant inherited AQP2 mutation were reported that do not lead to a retention in the Golgi apparatus. The C-terminal deletion mutants 727ΔG (Marr et al. 2000b) and 763 to 772del (Asai et al. 2003) are directed to the basolateral rather than to the correct apical membrane of the epithelial cells, suggesting that a post Golgi sorting step is affected.

Retinitis pigmentosa

Retinitis pigmentosa (RP) is a common inherited retinopathy affecting about 1 in 3,500 persons worldwide (van Soest et al. 1999; Chapple et al. 2001; Farrar et al. 2002). The disorder is associated with retinal atrophy, deposition of pigment and reduction of retinal vessels. At the beginning of the disease, the rods are the primarily affected photoreceptor cells, and clinical symptoms include a progressive loss of night vision as well as a bilateral symmetric loss of the mid-peripheral visual fields. With progression, cone photoreceptor cells are also affected, and day vision and central visual acuity are also compromised. The final stage of RP is severe visual impairment or blindness. The genetic and clinical heterogeneity of RP is remarkably high. Today, as many as 36 distinct gene loci associated with RP have been described (Phelan and Bok 2000; see also www.sph.uth.tmc.edu/retnet/disease.htm). So far, no clear genotype/phenotype relationship has been established. Mutations in the light receptor rhodopsin account for approximately 25% of all RP cases (Inglehearn et al. 1998). Since the first documentation of rhodopsin mutations associated with the disease (Dryja et al. 1990a; Dryja et al. 1990b), more than 150 different muta-

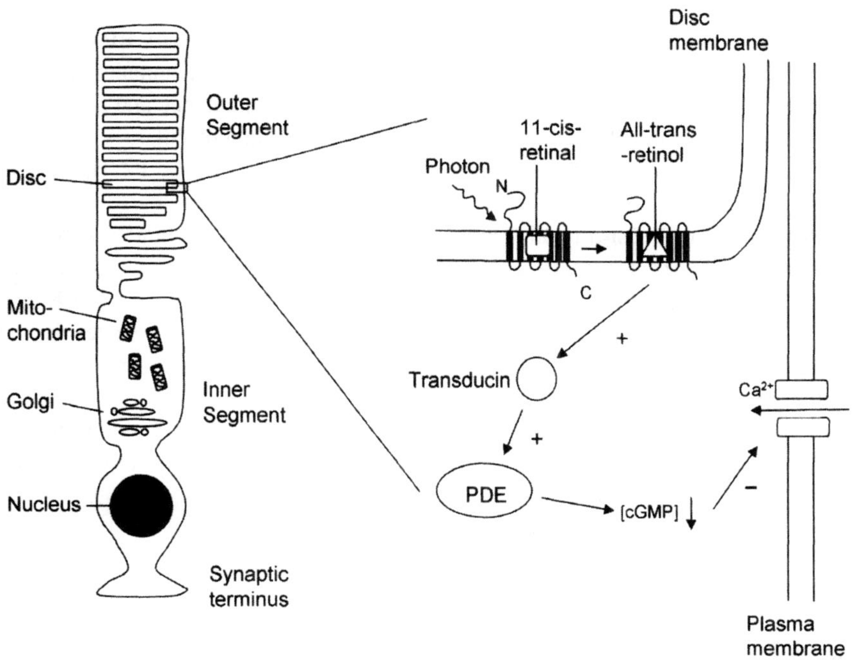

Fig. 8 Scheme of a rod photoreceptor cell (*left*) and the signal transduction initiated by rhodopsin (*right*). See text for details

tions have been described (Farrar et al. 2002; see also www.retina-international.com/sci-news/rhomut.htm).

Rhodopsin is a heptahelical type III membrane protein (349 amino acids) belonging to the GPCR protein family. It is expressed in rod photoreceptor cells. A rod cell (Fig. 8) consists of an outer segment containing the disc membranes, an inner segment and the synaptic terminus. While the apical part of the outer segment is embedded in the retinal epithelium the synaptic terminus connects the rods to the central nervous system (CNS). Rhodopsin is expressed in the disc membranes of the outer segment, contributing up to 85% of the total protein. The signal transduction pathway transforming incoming light to an electric signal is summarized in Fig. 8.

In contrast to the other GPCRs, the ligand of rhodopsin, the chromophore 11-*cis*-retinal, is bound by the receptor even in its inactive state. The chromophore is able to absorb a single photon. This causes the isomerization of 11-*cis* retinal to all-*trans*-retinol and a conformational change of the receptor leading to G protein (transducin) activation. Activated transducin stimulates the cGMP phosphodiesterase. The resulting decrease in cGMP leads to the closure of cGMP-gated cation channels at the plasma membrane. As a consequence, the plasma membrane hyperpolarizes, and the transmitter release at the synaptic terminus decreases.

Mutational loss of rhodopsin function explains RP symptoms such as night blindness and vision impairment. It does, however, not explain the dominant inheritance or the progressive retinal degradation. Interestingly, all RP-causing rhodopsin mutants seem to be impaired in intracellular trafficking. The transport of these mutants is difficult to assess. Transfected cells are not the best model, since the disc membranes representing the target compartment for rhodopsin are missing. Several transgenic animal models were developed to perform trafficking studies. In the rod cells of transgenic *Drosophila*, RP-causing rhodopsin mutants accumulated in the early secretory pathway (e.g., mutants G119E and P184L; Colley et al. 1995). It was speculated that dominant inheritance and retinal degeneration may result from ER overload with misfolded rhodopsin molecules. A similar cytotoxic effect was demonstrated for a rhodopsin triple mutant (V20G, P23H, P27L) expressed in a transgenic mouse line (Frederick et al. 2001).

In the rod cells of transgenic mice, transport-defective rhodopsin mutants were also localized in post ER compartments. The Q344ter mutant accumulated in the plasma membrane of the inner segment rather than in the disc membranes (Sung et al. 1994). The P347S mutant accumulated in extracellular vesicles near the junction between inner and outer segments (Li et al. 1996), and it was speculated that rod cell degeneration is caused by a failure to renew the outer segments at a normal rate.

Among the methods used to study rhodopsin trafficking, the recently published transgenic *Xenopus laevis* model is of particular interest since it will allow real-time transport studies using GFP fusions (Tam et al. 2000; Moritz et al. 2001).

Sucrase-isomaltase deficiency

Congenital sucrose-isomaltase deficiency (CSID) is a common autosomal recessive disorder (incidence 1:3,000) characterized by a defect in disaccharide absorption in the small intestine due to the absence of sucrase and maltase digestive activity (Treem 1995). Symptoms become evident in infants after disaccharide or oligosaccharide ingestion (such as sucrose or starch). Breast-fed infants or those on lactose formula only are not affected. Clinical signs and symptoms include an osmotic-fermentative diarrhea with dehydration, abdominal swelling and pain, and malnutrition with delay in growth and weight gain. The symptoms usually improve with age.

Mutations causing CSID are located in the gene for the enzyme sucrase-isomaltase (SI; Naim et al. 1988a). SI is a type II membrane protein that is exclusively expressed in the apical microvillar membranes of small intestine epithelial cells (Gorvel et al. 1991). The protein consists of 1,827 amino acids. It contains a short cytoplasmic N tail, a large extracellular C tail, and a single TM. The C tail contains a serine/threonine-rich domain thought to form a stalk for the two catalytical domains, the sucrase and isomaltase moieties. The two catalytical domains are proteolytically separated in the small intestine, but they remain associated through noncovalent interactions (Hauri et al. 1979; Naim et al. 1988b).

Many CSID-causing mutations lead to transport-defective enzymes. Interestingly, the mutant proteins are retained in different compartments along the secretory pathway: the Q1098P mutant is located in the ER but it is also able to reach the ERGIC (Ouwendijk et al. 1996; Moolenaar et al. 1997). Other mutants were detected even in the *medial* and *trans* Golgi compartments (Naim et al. 1988a), suggesting that the Golgi apparatus is involved in SI quality control. In addition, mutants sorting basolaterally rather than apically in epithelial cells have been found (Fransen et al. 1991; Spodsberg et al. 2001). Even the phenotype of a secreted rather than a membrane-anchored isomaltase moiety was described (Jacob et al. 2000).

The mechanisms and the proteins contributing to SI quality control have not been characterized yet. Studies addressing these questions are of particular interest, since SI quality control obviously also takes place in the Golgi apparatus. To date, quality control in the Golgi apparatus has been observed for only one other membrane protein, the AQP2 water channel (see "Congenital nephrogenic diabetes insipidus").

Familial hypercholesterolemia

Familial hypercholesterolemia (FH) is an autosomal dominant disease of lipoprotein metabolism (Nimpf and Schneider 2000; Smilde et al. 2001). Patients have very high levels of LDL-cholesterol. The consequences are early atherosclerosis and a high risk for myocardial infarction. In affected men, the heart attacks occur typically in the 40s to 50s. By age 60, 85% of men with FH have experienced a myocardial infarction. The incidence of heart attacks in affected women is also increased, but delayed 10 years in comparison to men. In homozygous patients, the severity of the condition is magnified. Affected individuals develop deposits of cholesterol beneath the skin over their elbows and knees (xanthomas) and around the cornea. Atherosclerosis begins typically before puberty, and myocardial infarctions and death may occur before age 30. Familial hypercholesterolemia is a common disorder with an estimated incidence of about 1 in 500.

Familial hypercholesterolemia is caused by mutations in the gene encoding the LDL receptor (Goldstein et al. 1977; Goldstein and Brown 1979). The LDL receptor is a type III membrane protein consisting of 839 amino acids. It internalizes circulating LDL particles by receptor-mediated endocytosis. The LDL receptor thus functions as an interface between extra- and intracellular cholesterol pools, removing cholesterol from the blood. Its loss of function in FH thus explains the hypercholesterolemia, the main symptom of the disease. The LDL receptor consists of five domains: (a) the ligand-binding domain interacting with LDL, (b) the epidermal growth factor homology domain playing a role in acidic-dependent LDL dissociation after internalization, (c) the O-linked sugar domain with unknown function, (d) the single TM anchoring the protein, and (e) the cytoplasmic domain triggering receptor-mediated endocytosis by a tyrosine-type endocytosis signal.

More than 500 different mutations have been identified in the LDL receptor gene (database www.ucl.ac.uk/fh/). It was estimated that about 50% of these mutations lead to intracellularly retained receptors (Hobbs et al. 1990). A recent study demonstrates that misfolded receptors display a prolonged association with the BiP chaperone (Jørgensen et al. 2000). Interestingly, no association with CNX/CRT was observed in this study although the LDL receptor is a glycoprotein.

Hereditary hemochromatosis

Hereditary hemochromatosis (HH) is a common autosomal recessive disorder of iron metabolism (Fleming and Sly 2002). The disease is characterized by abnormal, excessive iron absorption leading to saturation of serum transferrin and iron deposition in various cell types. The clinical consequences are multiorgan dysfunctions including hepatic failure, hepatic carcinoma, diabetes, cardiac failure, and arthritis.

Mutations causing HH are located in the gene encoding the human hemochromatosis protein (HFE), a 343 amino acid, major histocompatibility complex (MHC) class I-like protein (Feder et al. 1996). HFE is a type I membrane protein containing a cleavable signal peptide, a long extracellular N tail, a single TM and a relatively short cytoplasmic C tail. The protein forms a complex with β_2 microglobulin (Feder et al. 1996). Furthermore, it is associated with the transferrin receptor (Waheed et al. 1999). The HFE-containing complex is involved in sensing the body iron status in the crypt cells of the duodenum. Loss of function of the HFE protein leads to increased expression of genes involved in iron absorption by an as yet incompletely understood mechanism.

The vast majority of HH patients carry the C282Y mutation (United States and Europe=82–90%; Merryweather-Clarke et al. 1997). The resulting HFE protein is unable to bind β_2 microglobulin (Waheed et al. 1997). The unassembled form is recognized by the quality control system, retained intracellularly and finally subjected to proteolysis (Waheed et al. 1997). The intracellular retention of the C282Y mutant was not only observed for the recombinant protein expressed in COS.M6 cells, but also for the native protein in tissue samples of HH patients (granular staining in duodenal enterocytes; Zuccon et al. 2000). The prevalence of the C282Y mutation is remarkably high; the carrier frequency is highest among Caucasian individuals of European ancestry and reaches 10–15% (Beckman et al. 1997; Olynyk et al. 1999). Up to 0.5% of these populations are thus homozygous. However, not all individuals homozygous for the C282Y mutation develop clinical symptoms, suggesting that other proteins contribute to the clinical phenotype.

It is noteworthy that HH is associated with high morbidity and mortality but can easily be prevented by early identification and treatment. Because of its high prevalence, HH belongs to the group of diseases that seem to be suitable for genetic population screening. The future will show whether the consideration of all social and individual psychological consequences may allow such genetic screening.

Oculocutaneous albinism

Oculocutaneous albinism (OCA) is an autosomal recessive disease (incidence 1:15,000) characterized by a reduction or absence of melanin pigment in the skin, hair, and eyes. The pigment reduction in hair and skin leads to a change in color but not to an abnormal tissue development. The pigment reduction in the eye, however, causes a lack of formation of the fovea and misrouting of the optic nerves at the chiasm. Clinical symptoms include nystagmus, strabismus, and a decrease in visual acuity.

Mutations causing OCA type 1 (OCA1) are located in the gene for the enzyme tyrosinase (Tyr; Spritz et al. 1990; Oetting and King 1999), a copper-containing type I membrane protein consisting of 529 amino acids. The enzyme has a cleavable signal peptide, a long N tail, a single TM and a relatively short cytoplasmic C terminal domain. It is expressed in specialized compartments of melanocytes, the melanosomes, and catalyzes key reactions in melanin biosynthesis: (a) the hydroxylation of tyrosine to DOPA, (b) the oxidation of DOPA to dopaquinone, and (c) the oxidation of 5,6-dihydroxyindole to indole-5,6-quinone. Loss of tyrosinase function consequently leads to the absence of melanin pigment.

To date, 113 different mutant OCA1 alleles have been reported (database of the International Albinism Center at the University of Minnesota, www.cbc.umn.edu/iac). Individuals with OCA1A mutations produce completely inactive tyrosinases (classical form), whereas patients with OCA1B mutations have enzymes with some residual activity. The variation in the pigmentation defects resulting from OCA1B mutations ranges from very little to nearly normal skin and hair pigment. It was recently shown that at least some of the OCA1 mutations lead to Tyr enzymes that are retained in the early secretory pathway of melanocytes, namely the frequent T373 K mutation and the R402Q mutation (Halaban et al. 2000). The mutant proteins displayed a prolonged association with the lectine chaperones CNX and CRT (Halaban et al. 2000). The BiP chaperone also appears to play a role in this process (Toyofuku et al. 2001a). Studies with COS7 cells indicate that many other OCA1 mutations also lead to intracellularly retained Tyr enzymes (Toyofuku et al. 2001b). It is controversial, however, whether COS7 cells are a suitable model for Tyr transport since recent studies indicate that Tyr maturation within the ER requires melanocyte-specific factors (Francis et al. 2003).

The tyrosinase-related protein-1 (Tyrp1) is involved in melanin synthesis, too. Mutation of Tyrp1 cause OCA3 (Boissy et al. 1996). A retention of the mutant proteins in the early secretory pathway is discussed (Toyofuku et al. 2001a). The precise functions of two other proteins causing OCA are currently unknown: mutants of the P protein cause OCA2 (Durham-Pierre et al. 1994) and mutants of the membrane-associated transporter protein (MATP) cause OCA4 (Newton et al. 2001).

Laron syndrome

Laron syndrome (LS), also known as growth hormone insensitivity syndrome (GHIS), is a very rare autosomal recessive disorder causing dwarfism that develops shortly after birth (Hull and Harvey 1999; Savage et al. 2001). Worldwide, only about 250 patients are known. In addition to a short stature, patients may have deep-set eyes, small hands and feet, a high-pitched voice, and may be prone to obesity. Impaired intellectual development also appears to be associated with the disease. LS is an example of an inherited disorder with a considerable phenotype variation (Woods et al. 1997).

LS is caused by mutations in the gene for the growth hormone (GH) receptor (Godowski et al. 1989; Amselem et al. 1989) or sometimes from mutations in genes encoding proteins of the GH receptor signaling pathways (Laron et al. 1993). Patients with LS have normal GH levels. However, due to the lack of the GH receptor, the hormone is unable to exert its growth-promoting functions. Treatment of LS patients with GH is thus ineffective.

The GH receptor belongs to the large family of cytokine receptors. It has a wide tissue distribution but is particularly highly expressed in the liver. The GH receptor is a type 1 membrane protein of 619 amino acids containing a cleavable signal peptide, an N tail that forms the GH-binding domain, a single TM, and a cytoplasmic domain that couples to the cytosolic janus kinase (JAK) for signaling. In the liver, GH receptor activation leads to the secretion of insulin-like growth factor-1 (IGF-I) mediating bone growth. The ligand-engaged GH receptor is believed to form dimers. Two extracellular receptor domains appear to bind a single GH molecule (Frank 2002).

More than 30 different mutations have been identified in the GH receptor gene (Hull and Harvey 1999). At least three of them, I153T, Q154P, and V155G, lead to receptors that are retained intracellularly (Wojcik et al. 1998). The folding defects emerging from these mutations can be derived from structural data available for the GH receptor. The crystal structure of the extracellular domain (De Vos et al. 1992) indicates that residues Ile^{153} and V^{155} stabilize a hydrophobic core ("topohydrophobic residues", Wojcik et al. 1998). The decrease in hydrophobicity resulting from the I153T and V154G mutations seems to destabilize this hydrophobic core and leads to a prolonged association of the GH receptor mutants with so far unknown ER chaperones.

Other LS mutations cause LH binding defects. Since the intracellular transport of the corresponding receptor mutants was not assessed in detail, it is unclear whether these binding defects are due to changes in the binding properties of the receptor exposed on the cell surface or to transport defects. The number of transport-deficient LH receptor mutants may thus be higher than currently estimated.

Congenital long QT syndrome

The congenital long QT syndrome (CLQTS) is a disease predisposing to cardiac arrhythmia (Ashcroft 2000; Splawski et al. 2000; Towbin et al. 2001). The term "long QT syndrome" refers to an abnormality found in the electrocardiogram: a long QT interval caused by a prolonged repolarization period of the cardiomyocytes after an action potential. In general, CLQTS patients present with a specific type of episodic arrhythmia known as *torsade de pointes*, so called because the QRS waves of the electrocardiogram appear as oscillations around the isoelectric line. The rapid contractions associated with *torsade de pointes* arrhythmia lead to dramatically reduced ventricular filling and consequently to a sudden loss of consciousness. More importantly, this type of arrhythmia may cause sudden cardiac death by developing into ventricular fibrillation.

The occurrence of this type of arrhythmia is usually provoked by stressors ranging from physical exercise to loud sounds. The exact number of affected patients is unknown,but it is estimated that in the USA over 50,000 people suffer from CLQTS (prevalence 1:5,000), and that the disease may cause as many as 3,000 deaths each year (mostly of children and young adults). Mutations causing CLQTS were found in genes encoding subunits of ion channels involved in depolarization and repolarization of cardiac myocytes.

IKr potassium channel mutations. The voltage-gated cardiac potassium channel mediating the I_{Kr} current (rapidly activating delayed rectifier potassium current) consists of α and β subunits (Fig. 9). The α subunit is encoded by the gene for the human ether-a-go-go-related protein (HERG; also known as KCNH2, LQT2). The β subunit is encoded by the gene for the minK-related peptide 1 (MiRP1; also known as KCNE2). Four HERG molecules associate with an unknown number of MiRP1 molecules to establish active I_{Kr} channels. It is believed that the four α subunits assemble in a four-fold symmetry to form the central ion permeation pore.

The HERG protein is a hexahelical type II membrane protein consisting of 1,159 amino acids (Fig. 9) with the typical topology and domain organization of voltage-gated potassium channels: an amphipathic TM4 with charged residues believed to form the voltage sensor, a TM5-TM6 connecting loop involved in pore wall formation, and a TM4-TM5 connecting loop involved in channel inactivation.

Mutations in the HERG gene cause an autosomal dominant form of CLQTS (Curran et al. 1995). The prolonged QT interval is explicable assuming that the mutations lead to nonfunctional HERG proteins. In the normal cardiac action potential, the depolarizing inward current through sodium channels is not immediately reversed by an outward repolarizing current, as in neurons. The repolarizing current through three types of potassium channels (I_{Kr}, I_{Ks} and I_{Kur} channels, "delayed rectifiers") is initially balanced by a depolarizing current through calcium channels. Mutations in the HERG gene impair the I_{Kr} current. This leads to an even longer repolarization period and a prolonged QT interval.

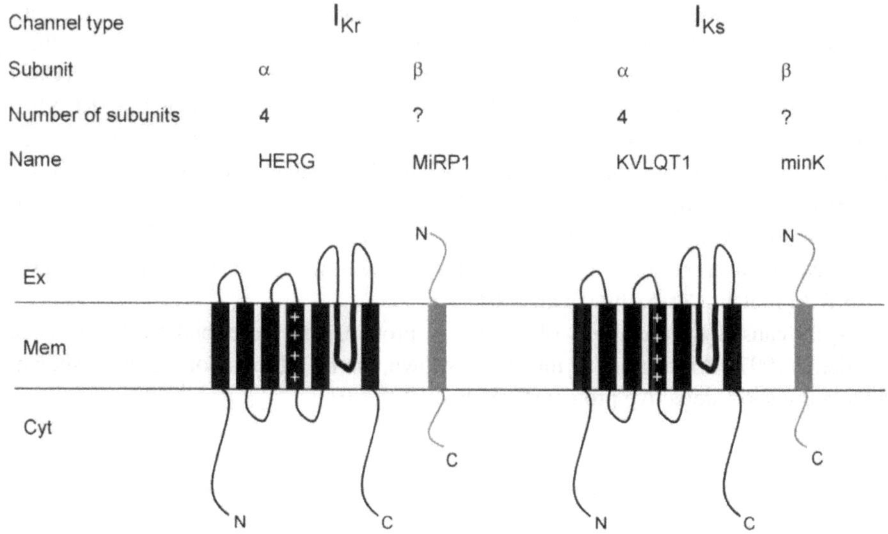

Fig. 9 Subunit composition, topology and functional domains of the cardiac voltage-gated potassium channels I_{Kr} (*left*) and I_{Ks} (*right*) affected by CLQTS mutations. The α and β subunits are indicated in *black* and *gray* respectively. The voltage sensor in TM4 of each α subunit is indicated by "+" symbols, the TM5-TM6 connecting loop forming the ion pore wall is indicated by a *heavy line*. Both β subunits contain a cleavable signal peptide (not shown). See text for further details. *Ex* extracellular space, *Mem* plasma membrane, *Cyt*=cytoplasm

More than 90 different HERG mutations are known (Splawski et al. 2000). At least some of them lead to transport-deficient proteins. Among these are the completely transport-deficient mutants Y611H and V822M (Zhou et al. 1998), A561V (Ficker et al. 2000a), R752W (Ficker et al. 2000b), and T65P (Paulussen et al. 2002). The G601S mutant, in contrast, displayed only a partial transport defect (Furutani et al. 1999). Trafficking of the HERG protein is not very well understood. It was shown that the cytosolic chaperones Hsp70 and Hsp90, but not Grp94, are involved in the quality control of the protein in the ER (Ficker et al. 2003). A nine amino acid residue segment in the C-tail of the protein seems to be required for establishing a transport-competent folding state (Akhavan et al. 2003). HERG trafficking in the Golgi apparatus seems to be influenced by the Golgi matrix protein 130 (GM130) by an as yet unidentified mechanism (Roti et al. 2002). The transport defects of the HERG mutants together with the fact that the protein forms tetramers explains why CLQTS is transmitted as an autosomal dominant trait. In the heterozygous situation, mutant and wild-type proteins coexist. However, if subunit assembly takes place in the ER, the combination of wild-type and mutant subunits may lead to the retention of the whole channel complex (Ficker et al. 2000a).

The β subunit of the I_{Kr} channel, the MiRP1 protein, is a small type I membrane protein (123 amino acids) containing a single TM, an extracellular N tail and an intracellular C tail (Fig. 9; Abbott et al. 1999). The precise function of MiRP1 within the channel complex is unknown. CLQTS-causing mutations were also described for this protein (Q9E, M54T, I57T; Abbott et al. 1999), but intracellular trafficking has not been studied thus far.

IKs potassium channel mutations. The voltage-gated cardiac potassium channel mediating the I_{Ks} current (slowly activating delayed rectifier potassium current) is similar to the I_{Kr} channel (Fig. 9). The α subunit is encoded by the KVLQT1 gene (also known as KCNQ1, LQT1, CIQ1). The β subunit is encoded by the minK gene (also known as ISK, KCNE1).

The mechanisms by which inactivating mutations in KVLQT1 and minK provoke a prolonged QT interval are also equivalent since the I_{Ks} channel is likewise involved in repolarization of the cardiac action potential. Mutations in the KVLQT1 gene were originally described by Wang and coworkers (1996). To date, more than 100 different mutations are known. A transport defect resulting from misfolding was recently found in the case of the T587 M mutant (Yamashita et al. 2001).

CLQTS caused by mutations of the minK protein were described by Splawski and coworkers (1997). Among the 12 mutations known, the L51H mutation has been shown to lead to a transport-defective protein (Bianchi et al. 1999).

Charcot-Marie-Tooth syndrome

Charcot-Marie-Tooth syndrome (CMT) is a common, inherited neurological disease characterized by the demyelination of peripheral neurons (incidence=1:2,500; Sanders et al. 2001; Young and Suter 2001). The disorder presents with a slowly progressive degeneration of the muscles in the foot, lower leg, hand, and forearm. Other symptoms include a mild loss of sensation in the limbs, fingers, and toes. Foot bone abnormalities such as high arches and hammer toes may occur. Full expression of CMT's clinical symptoms generally occurs by age 30.

Genetically, CMT is a heterogeneous disorder. The most common type, CMT1A, is inherited as an autosomal dominant trait. It is caused by mutations in the gene for the peripheral myelin protein 22 (PMP22), a small dimeric tetrahelical type II membrane protein consisting of 160 amino acids. PMP22 it is a critical component of the mammalian myelin sheath, comprising up to 2–5% of the total protein. The myelin sheath is generated by Schwann cells which radiate and wrap around the axons of peripheral neurons. Compact myelin consists of alternating layers of plasma membranes of Schwann cells separated by thin layers of Schwann cell cytoplasm or ectoplasm. The PMP22 protein is located in the membranes of these layers. It associates with the P_0 glycoprotein, which connects the membrane layers by homophilic interactions. PMP22 is believed to assist in the gluing functions of P_0. This would explain the demyelinating phenotype associated with PMP22 mutations.

Interestingly, the complete deletion of the PMP22 gene leads to a very mild form of the disease while most missense mutations cause severe phenotypes (Sanders et al. 2001). This paradox was solved by investigating the intracellular transport of the PMP22 proteins carrying missense mutations. The vast majority of the PMP22 missense mutants were retained in the ER (Naef et al. 1997; Naef and Suter 1999; Colby et al. 2000). These mutants not only fail to exit the early secretory pathway, they also dramatically decrease the transport of coexpressed wild-type proteins (Tobler et al. 1999). These data are plausible assuming that PMP22 dimerizes in the early secretory pathway (Ryan et al. 2000) and that the missense mutants keep their ability to dimerize but fail to pass the quality control system. Such a mechanism is reminiscent of that described for HERG mutants causing congenital long QT syndrome (see "Congenital long QT syndrome"). A dimerization of PMP22 in the early secretory pathway not only explains why CMT1A is transmitted as an autosomal dominant trait, but also why a complete deletion of the PMP22 gene causes only a mild phenotype in a heterozygous situation. In this case, trafficking of wild-type PMP22 is not affected.

Mutations of other tetrahelical membrane proteins that are structurally and/or functionally related to PMP22 are also associated with diseases (Sanders et al. 2001). The tight junction protein paracellin 1 (claudin 16), whose mutants are associated with hypomagnesemia (Simon et al. 1999), and the calcium channel γ subunit stargazin, whose mutants are associated with murine epilepsy (Letts et al. 1998) belong to this group. Trafficking defects in the early secretory pathway have been described for mutants of the proteolipid protein (PLP) associated with the dysmyelinating Pelizaeus-Merzbacher disease (Gow et al. 1994; Thomson et al. 1997; Gow et al. 1998; Southwood and Gow 2001). The same holds true for mutants of the connexin (Cx) proteins of gap junctions like the Cx26 mutants associated with hereditary deafness (Martin et al. 1999; White 2000) and the Cx32 mutants associated with an X-linked form of the Charcot-Marie-Tooth syndrome (Deschenes et al. 1997; Bone et al. 1997; Martin et al. 2000).

Pharmacological approaches to correct membrane protein misfolding and retention in the early secretory pathway

Many of the diseases caused by misfolded membrane proteins have a poor prognosis and their symptomatic therapy is inefficient (e.g., cystic fibrosis). If a mutation in a membrane protein leads to a misfolded and transport-defective form, the development of drugs to overcome this situation appears to be difficult. However, recently performed in vitro studies to correct misfolding of membrane proteins are promising.

Two pharmacological strategies are being pursued at the moment. The first strategy involves the development of substances that favor correct folding of the mutant proteins and consequently allow them to pass the quality control system. The second strategy is based on the fact that the quality control system is sometimes overprotective and retains mutant proteins that are still functional. Here, substances that do not influence folding but prevent the interaction with quality control components may allow transport and lead to a functional rescue. It is conceivable that a combination of a drug favoring correct folding with a drug inhibiting the interaction with the quality control system will be most effective for the treatment of diseases caused by transport-defective proteins. Most studies have been carried out with the ΔF508 mutant of the CFTR protein.

Development of substances that favor correct membrane protein folding

Chemical chaperones. The observation that a reduction in temperature favors correct folding of the ΔF508 CFTR mutant (Denning et al. 1992; Brown et al. 1997) led to the idea that this effect may also be achieved by the use of substances acting as "chemical chaperones," i.e., by substances improving protein folding by nonspecific interactions. Indeed, it was shown that the amount of correctly processed and functional ΔF508 CFTR protein was increased if cells were treated with glycerol (Sato et al. 1996) or trimethylamine-*N*-oxide (Brown et al. 1996), although the rescued channels showed a decreased half-life under these conditions (Sharma et al. 2001). Another substance restoring some cell surface transport of the ΔF508 mutant is 4-phenylbutyrate (Rubenstein et al. 1997; Loffing et al. 1999; Moyer et al. 1999). The nonspecific action of chemical chaperones is reflected by the fact that they were able to rescue many other folding-defective mutant proteins. Glycerol, for example, was shown to rescue AQP2 mutants (Tamarappoo et al. 1999) and the Z $\alpha 1$-antitrypsin mutant found in $\alpha 1$ antitrypsin deficiency (Burrows et al. 2000).

Pharmacological chaperones. The most important disadvantage of chemical chaperones is that they must be used at very high concentrations, precluding an in vivo application. This led to the idea that correct folding states may also be stabilized by compounds binding specifically to target proteins. For example, addition of a ligand during the folding process of a mutant receptor may help to establish and maintain the correct conformation of the ligand-binding pocket and consequently the correct structure of the whole protein. If such "pharmacological chaperones" interact with the extracellular domains of membrane proteins, they must be hydrophobic enough to pass not only the plasma membrane but also the ER membrane.

In the case of the ΔF505 CFTR mutant, the A_1 adenosine receptor antagonist 8-cyclopentyl-1,2-dipropylxanthine (CPX) is such a compound that seems to exhibit a more specific interaction. It binds with high affinity to the first nucleotide-binding domain of CFTR, which structurally resembles an A_1 adenosine receptor binding site (Cohen et al. 1997). In transfected cells, CPX leads to a restoration of the plasma membrane transport of the ΔF508 CFTR mutant, most likely by stabilizing correct folding (Zeitlin 2000). For this compound, phase II clinical trials with cystic fibrosis patients are underway. Likewise, the CFTR-activating benzo(c)quinolizinium drugs MPB-07 and MPB-91 restored the transport of the ΔF508 CFTR mutant in vitro. This effect, too, appears to involve a specific binding site (Dormer et al. 2001). However, experiments directly determining the effect of the drugs on ΔF508 CFTR folding or exit from the early secretory pathway are needed to justify the conclusion that these drugs favor or improve correct folding of the mutant protein.

The concept of pharmacological chaperones was recently extended to proteins unrelated to CFTR. By using the selective nonpeptidic V_2R antagonist SR121463A, the function of 8 NDI-causing V_2R mutants could be rescued in transfected cells by promoting their correct folding (Morello et al. 2000a; Morello et al. 2000b). Moreover, the retinitis pigmentosa-causing P23H mutant of rhodopsin could be rescued by the retinal derivative 11-*cis*-7-ring retinal (Noorwez et al. 2003). Similarly, the cell surface transport of the long QT syndrome-causing HERG mutants N470D and G601S could be rescued by antiarrhythmic drugs that normally block HERG channels (Zhou et al. 1999; Ficker et al. 2002).

The pharmacological chaperone action appears to be independent of the intrinsic signaling efficacy of the ligand. In the case of the delta opioid receptor, both agonists and antagonists were found to promote correct receptor folding (Petaja-Repo et al. 2002).

Development of substances that prevent the interaction of components of the quality control system with target proteins

Neither chemical nor pharmacological chaperones lead to wild-type expression levels of the mutant proteins at the cell surface. Alternative or additional strategies are needed to improve the intracellular transport of the mutant proteins. In the future, drugs may be developed that influence components of the quality control system involved in the retention of misfolded proteins. Inhibitors of quality control components may be combined with pharmacological chaperones: the pharmacological chaperones initially promote correct folding of a mutant protein, the quality control inhibitors then improve its cell surface transport. The latter compounds would also be useful for mutant proteins that are still functional but nevertheless retained by the quality control system, i.e., in cases when the quality control system is overprotective. Examples include the ΔF508 mutant of the CFTR protein (Pasyk and Foskett 1995) and the T126M mutant of the AQP2 protein (Tamarappoo and Verkman 1998; Tamarappoo et al. 1999).

Promising new drug targets are the adapter proteins recognizing ER retention signals exposed by at least some misfolded proteins (signal-based quality control, see "Signal-based quality control"). If the misfolded proteins reach the ERGIC, the proteins mediating the retrograde transport back to the ER may also be included. It was postulated that the retrograde transport is initiated by association of the misfolded protein with BiP (Yamamoto et al. 2001). BiP contains a C-terminal KDEL retrieval signal that enables the BiP/mutant protein complex to enter the retrograde transport vesicles by binding to the KDEL receptor (Yamamoto et al. 2001; see the section entitled "Quality control in the ERGIC").

Drugs inhibiting the KDEL receptor are so far not known. BiP is a chaperone protein containing a substrate polypeptide binding site and an ATP binding site. It was recently shown that the anti-inflammatory drug salicylic acid binds specifically to the polypeptide binding site of BiP in human cells, and it was proposed that inhibition of BiP may interfere with folding and transport of inflammatory proteins (Deng et al. 2001). The salicylates may thus represent key compounds for the development of substances that inhibit BiP and rescue misfolded proteins retained in the ERGIC.

Acknowledgments. The author would like to thank Walter Rosenthal, Ricardo Hermosilla and Alexander Oksche for critical readings of the manuscript and for useful discussions. Our own work in the field was supported by the Deutsche Forschungsgemeinschaft (SFB 249, SFB 366, SFB 449).

References

Abbott GW, Sesti F, Splawski I, Buck ME, Lehmann MH, Timothy KW, Keating MT, Goldstein SA (1999) MiRP1 forms I_{Kr} potassium channels with HERG and is associated with cardiac arrhythmia Cell 97:175–187

Akhavan A, Atanasiu R, Shrier A (2003) Identification of a COOH-terminal segment involved in maturation and stability of human ether-a-go-go-related gene potassium channels. J Biol Chem 278:40105–40112

Ala Y, Morin D, Mouillac B, Sabatier N, Vargas R, Cotte N, Dechaux M, Antignac C, Arthus MF, Lonergan M, Turner MS, Balestre MN, Alonso G, Hibert M, Barberis C, Hendy GN, Bichet DG, Jard S (1998) Functional studies of twelve mutant V2 vasopressin receptors related to nephrogenic diabetes insipidus: molecular basis of a mild clinical phenotype. J Am Soc Nephrol 9:1861–1872

Amselem S, Duquesnoy P, Attree O, Novelli G, Bousnina S, Postel-Vinay MC, Goossens M (1989) Laron dwarfism and mutations of the growth hormone-receptor gene. N Engl J Med 321:989–995

Andersen-Beckh B, Dehe M, Schülein R, Wiesner B, Rutz C, Liebenhoff U, Rosenthal W, Oksche A (1999) Polarized expression of the vasopressin V2 receptor in Madin-Darby canine kidney cells. Kidney Int. 56:517–527

Andersson H, von Heijne G (1994) Membrane protein topology: effects of $\Delta\mu H+$ on the translocation of charged residues explain the 'positive inside' rule. EMBO J 13:2267–2272

Andersson H, D'Antona AM, Kendall DA, von Heijne G, Chin CN (2003) Membrane assembly of the cannabinoid receptor 1: impact of a long N-terminal tail. Mol Pharmacol 64:570–577

Antonny B, Schekman R (2001) ER export: public transportation by the COPII coach. Curr Opin Cell Biol 13:438–443

Appenzeller C, Andersson H, Kappeler F, Hauri HP (1999) The lectin ERGIC-53 is a cargo transport receptor for glycoproteins. Nat Cell Biol 1:330–334

Aridor M, Bannykh SI, Rowe T, Balch WE (1995) Sequential coupling between COPII and COPI vesicle coats in endoplasmic reticulum to Golgi transport. J Cell Biol 131:875–893

Aridor M, Hannanä LA (2000) Traffic jam: a compendium of human diseases that affect intracellular transport processes. Traffic 1:836–851

Aridor M, Fish KN, Bannykh S, Weissman J, Roberts TH, Lippincott-Schwartz J, Balch WE (2001) The Sar1 GTPase coordinates biosynthetic cargo selection with endoplasmic reticulum export site assembly. J Cell Biol 152:213–229

Asai T, Kuwahara M, Kurihara H, Sakai T, Terada Y, Marumo F, Sasaki S (2003) Pathogenesis of nephrogenic diabetes insipidus by aquaporin-2 C-terminus mutations. Kidney Int 64:2–10

Ashcroft FM (2000) Ion channels and disease, Academic Press, San Diego, USA

Bannykh SI, Nishimura N, Balch WE (1998) Getting into the Golgi. Trends Cell Biol 8:21–25

Bannykh SI, Bannykh GI, Fish KN, Moyer BD, Riordan JR, Balch WE (2000) Traffic pattern of cystic fibrosis transmembrane regulator through the early exocytic pathway. Traffic 1:852–780

Beckman LE, Saha N, Spitsyn V, Van Landeghem G, Beckman L (1997) Ethnic differences in the HFE codon 282 (Cys/Tyr) polymorphism. Hum Hered 47:263–267

Belden WJ, Barlowe C (2001) Role of Erv29p in collecting soluble secretory proteins into ER-derived transport vesicles. Science 294:1528–1531

Bermak JC, Li M, Bullock C, Zhou QY (2001) Regulation of transport of the dopamine D1 receptor by a new membrane-associated ER protein. Nat Cell Biol 3:492–498

Bianchi L, Shen Z, Dennis AT, Priori SG, Napolitano C, Ronchetti E, Bryskin R, Schwartz PJ, Brown AM (1999) Cellular dysfunction of LQT5-minK mutants: abnormalities of I_{Ks}, I_{Kr} and trafficking in long QT syndrome. Hum Mol Genet 8:1499–1507

Bockaert J, Pin JP (1999) Molecular tinkering of G protein-coupled receptors: an evolutionary success. EMBO J 18:1723–1729

Bockenhauer D (2001) Ion channels in disease. Curr Opin Pediatr 13:142–149

Boissy RE, Zhao H, Oetting WS, Austin LA, Wildenberg SC, Boissy Y, Sturm RA, Hearing VJ, King RA, Nordlund JJ (1996) Mutation in and lack of expression of tyrosinase-related protein-1 (TRP-1) in melanocytes from an individual with brown oculocutaneous albinism: a new subtype of albinism classified as "OCA3". Am J Hum Gen 58:1145–1156

Bone LJ, Deschenes SM, Balice-Gordon RJ, Fischbeck KH, Scherer SS (1997) Connexin32 and X-linked Charcot-Marie-Tooth disease. Neurobiol Dis 4:221–230

Borel AC, Simon SM (1996) Biogenesis of polytopic membrane proteins: membrane segments assemble within translocation channels prior to membrane integration. Cell 85:379–389

Bonifacino JS, Weissman AM (1998) Ubiquitin and the control of protein fate in the secretory and endocytic pathways. Annu Rev Cell Dev Biol 14:19–57

Bradbury NA (1999) Intracellular CFTR: localization and function. Physiol Rev 79:S175–191

Brodsky JL (1998) Translocation of proteins across the endoplasmic reticulum membrane. Int Rev Cytol 178:277–328

Brown CR, Hong-Brown LQ, Biwersi J, Verkman AS, Welsh WJ (1996) Chemical chaperones correct the mutant phenotype of the ΔF508 cystic fibrosis transmembrane conductance regulator protein. Cell Stress Chaperones 1:117–125

Brown CR, Hong-Brown LQ, Welch WJ (1997) Correcting temperature-sensitive protein folding defects. J Clin Invest 99:1432–1444

Burrows JA, Willis LK, Perlmutter DH (2000) Chemical chaperones mediate increased secretion of mutant α1-antitrypsin (α1-AT) Z: A potential pharmacological strategy for prevention of liver injury and emphysema in α1-AT deficiency. Proc Natl Acad Sci USA 97:1796–1801

Canfield MC, Tamarappoo BK, Moses AM, Verkman AS, Holtzman EJ (1997) Identification and characterization of aquaporin-2 water channel mutations causing nephrogenic diabetes insipidus with partial vasopressin response. Hum Mol Genet 6:1865–71

Cao G, Kuhn A, Dalbey RE (1995) The translocation of negatively charged residues across the membrane is driven by the electrochemical potential: evidence for an electrophoresis-like membrane transfer mechanism. EMBO J 14:866–875

Celesia GG (2001) Disorders of membrane channels or channelopathies. Clin Neurophysiol 112:2–18

Chang XB, Cui L, Hou YX, Jensen TJ, Aleksandrov AA, Mengos A, Riordan JR (1999) Removal of multiple arginine-framed trafficking signals overcomes misprocessing of ΔF508 CFTR present in most patients with cystic fibrosis. Mol Cell 4:137–142

Chapple JP, Grayson C, Hardcastle AJ, Saliba RS, van der Spuy J, Cheetham ME (2001) Unfolding retinal dystrophies: a role for molecular chaperones? Trends Mol Med 7:414–421
Cheng SH, Gregory RJ, Marshall J, Paul S, Souza DW, White GA, O'Riordan CR, Smith AE (1990) Defective intracellular transport and processing of CFTR is the molecular basis of most cystic fibrosis. Cell 63:827–834
Chin CN, von Heijne G, de Gier JW (2002) Membrane proteins: shaping up. Trends Biochem Sci 27:231–234
Cohen BE, Lee G, Jacobson KA, Kim YC, Huang Z, Sorscher EJ, Pollard HB (1997) 8-cyclopentyl-1,3-dipropylxanthine and other xanthines differentially bind to the wild-type and ΔF508 first nucleotide binding fold (NBF-1) domains of the cystic fibrosis transmembrane conductance regulator. Biochemistry 36:6455–6461
Colby J, Nicholson R, Dickson KM, Orfali W, Naef R, Suter U, Snipes GJ (2000) PMP22 carrying the trembler or trembler-J mutation is intracellularly retained in myelinating Schwann cells. Neurobiol Dis 7, 561–573
Colley NJ, Cassill JA, Baker EK, Zuker CS (1995) Defective intracellular transport is the molecular basis of rhodopsin-dependent dominant retinal degeneration. Proc Natl Acad Sci USA 92:3070–3074
Curran ME, Splawski I, Timothy KW, Vincent GM, Green ED, Keating MT (1995) A molecular basis for cardiac arrhythmia: HERG mutations cause long QT syndrome. Cell 80:795–803
Cyr DM, Hohfeld J, Patterson C (2002) Protein quality control: U-box-containing E3 ubiquitin ligases join the fold Trends Biochem. Sci 27:368–375
Danilczyk UG, Cohen-Doyle MF, Williams DB (2000) Functional relationship between calreticulin, calnexin, and the endoplasmic reticulum luminal domain of calnexin. J Biol Chem 275:13089–13097
De Vos AM, Ultsch M, Kossiakoff AA (1992) Human growth hormone and extracellular domain of its receptor: crystal structure of the complex. Science 255:306–312
Deen PM, Verdijk MA, Knoers NV, Wieringa B, Monnens LA, van Os CH, van Oost BA (1994) Requirement of human renal water channel aquaporin-2 for vasopressin-dependent concentration of urine. Science 264:92–95
Deen PM, Croes H, van Aubel RA, Ginsel LA, van Os CH (1995) Water channels encoded by mutant aquaporin-2 genes in nephrogenic diabetes insipidus are impaired in their cellular routing. J Clin Invest 95:2291–2296
Delgado-Partin VM, Dalbey RE (1998) The proton motive force, acting on acidic residues, promotes translocation of amino-terminal domains of membrane proteins when the hydrophobicity of the translocation signal is low. J Biol Chem 273:9927–9934
Deng WG, Ruan KH, Du M, Saunders MA, Wu KK (2001) Aspirin and salicylate bind to immunoglobulin heavy chain binding protein (BiP) and inhibit its ATPase activity in human fibroblasts. FASEB J 15:2463–2470
Denning GM, Anderson MP, Amara JF, Marshall J, Smith AE, Welsh MJ (1992) Processing of mutant cystic fibrosis transmembrane conductance regulator is temperature-sensitive. Nature 358:761–764
Deschenes SM, Walcott JL, Wexler TL, Scherer SS, Fischbeck KH (1997) Altered trafficking of mutant connexin32. J Neurosci 17:9077–9084
Do H, Falcone D, Lin J, Andrews DW, Johnson AE (1996) The cotranslational integration of membrane proteins into the phospholipid bilayer is a multistep process. Cell 85:369–78
Dormer RL, Derand R, McNeilly CM, Mettey Y, Bulteau-Pignoux L, Metaye T, Vierfond JM, Gray MA, Galietta LJ, Morris MR, Pereira MM, Doull IJ, Becq F, McPherson MA (2001) Correction of ΔF508-CFTR activity with benzo(c)quinolizinium compounds through facilitation of its processing in cystic fibrosis airway cells. J Cell Sci 114:4073–4081
Dryja TP, McGee TL, Reichel E, Hahn LB, Cowley GS, Yandell DW, Sandberg MA, Berson EL (1990a) A point mutation of the rhodopsin gene in one form of retinitis pigmentosa. Nature 343:364–366
Dryja TP, McGee TL, Hahn LB, Cowley GS, Olsson JE, Reichel E, Sandberg MA, Berson EL (1990b) Mutations within the rhodopsin gene in patients with autosomal dominant retinitis pigmentosa. N Engl J Med 323:1302–1307
Durham-Pierre D, Gardner JM, Nakatsu YU, King RA, Francke U, Ching A, Aquaron R, de Marmol V, Brilliant M H (1994) African origin of an intragenic deletion of the human P gene in tyrosinase positive oculocutaneous albinism. Nat Genetics 7:176–179
Edvardsen O, Reiersen AL, Beukers MW, Kristiansen K (2002) tGRAP, the G protein-coupled receptors mutant database. Nucleic Acids Res 30:361–363
Ellgaard L, Helenius A (2001) ER quality control: towards an understanding at the molecular level. Curr Opin Cell Biol 13:431–437
Ellgaard L, Riek R, Braun D, Herrmann T, Helenius A, Wüthrich K (2001a) Three-dimensional structure topology of the calreticulin P-domain based on NMR assignment. FEBS Lett 488:69–73

Ellgaard L, Riek R, Herrmann T, Guntert P, Braun D, Helenius A, Wüthrich K (2001b) NMR structure of the calreticulin P-domain. Proc Natl Acad Sci USA 98:3133–3138

Farrar GJ, Kenna PF, Humphries P (2002) On the genetics of retinitis pigmentosa and on mutation-independent approaches to therapeutic intervention. EMBO J 21:857–864

Feder JN, Gnirke A, Thomas W, Tsuchihashi Z, Ruddy DA, Basava A, Dormishian F, Domingo R Jr, Ellis MC, Fullan A, Hinton LM, Jones NL, Kimmel BE, Kronmal GS, Lauer P, Lee VK, Loeb DB, Mapa FA, McClelland E, Meyer NC, Mintier GA, Moeller N, Moore T, Morikang E, Wolff RK, et al (1996) A novel MHC class I-like gene is mutated in patients with hereditary haemochromatosis. Nat Genet 13:399–408

Ficker E, Dennis AT, Obejero-Paz CA, Castaldo P, Taglialatela M, Brown AM (2000a) Retention in the endoplasmic reticulum as a mechanism of dominant-negative current suppression in human long QT syndrome. J Mol Cell Cardiol 32:2327–2337

Ficker E, Thomas D, Viswanathan PC, Dennis AT, Priori SG, Napolitano C, Memmi M, Wible BA, Kaufman ES, Iyengar S, Schwartz PJ, Rudy Y, Brown AM (2000b) Novel characteristics of a misprocessed mutant HERG channel linked to hereditary long QT syndrome. Am J Physiol Heart Circ Physiol 279:H1748–1756

Ficker E, Obejero-Paz CA, Zhao S, Brown AM (2002) The binding site for channel blockers that rescue misprocessed human long QT syndrome type 2 ether-a-gogo-related gene (HERG) mutations. J Biol Chem 277:4989–4998

Ficker E, Dennis AT, Wang L, Brown AM (2003) Role of the cytosolic chaperones Hsp70 and Hsp90 in maturation of the cardiac potassium channel HERG. Circ Res 92:e87–100

Fiedler K, Parton RG, Kellner R, Etzold T, Simons K (1994) VIP36, a novel component of glycolipid rafts and exocytic carrier vesicles in epithelial cells. EMBO J 13:1729–1740

Fiedler K, Veit M, Stamnes MA, Rothman JE (1996) Bimodal interaction of coatomer with the p24 family of putative cargo receptors. Science 273:1396–1399

Fleming RE, Sly WS (2002) Mechanisms of iron accumulation in hereditary hemochromatosis Annu Rev Physiol. 64:663–680

Francis E, Wang N, Parag H, Halaban R, Hebert DN (2003) Tyrosinase maturation and oligomerization in the endoplasmic reticulum require a melanocyte-specific factor. J Biol Chem 278:25607–25617

Frank SJ (2002) Receptor dimerization in GH and erythropoietin action—it takes two to tango, but how? Endocrinology 143:2–10

Fransen JA, Hauri HP, Ginsel LA, Naim HY (1991) Naturally occurring mutations in intestinal sucrase-isomaltase provide evidence for the existence of an intracellular sorting signal in the isomaltase subunit. J Cell Biol 115:45–57

Frederick JM, Krasnoperova NV, Hoffmann K, Church-Kopish J, Ruther K, Howes K, Lem J, Baehr W (2001) Mutant rhodopsin transgene expression on a null background. Invest Ophthalmol Vis Sci 42:826–833

Fushimi K, Uchida S, Hara Y, Hirata Y, Marumo F, Sasaki S (1993) Cloning and expression of apical membrane water channel of rat kidney collecting tubule. Nature 361:549–552

Furutani M, Trudeau MC, Hagiwara N, Seki A, Gong Q, Zhou Z, Imamura S, Nagashima H, Kasanuki H, Takao A, Momma K, January CT, Robertson GA, Matsuoka R (1999) Novel mechanism associated with an inherited cardiac arrhythmia: defective protein trafficking by the mutant HERG (G601S) potassium channel. Circulation 99:2290–2294

Furman I, Cook O, Kasir J, Low W, Rahamimoff H (1995) The putative amino-terminal signal peptide of the cloned rat brain Na(+)-Ca2+ exchanger gene (RBE-1) is not mandatory for functional expression. J Biol Chem 270:19120–19127

Gething MJ (1999) Role and regulation of the ER chaperone BiP Semin Cell Dev Biol 10:465–72

Gilbert A, Jadot M, Leontieva E, Wattiaux-De Coninck S, Wattiaux R (1998) ΔF508 CFTR localizes in the endoplasmic reticulum-Golgi intermediate compartment in cystic fibrosis cells. Exp Cell Res 242:144–52

Gilmore R, Blobel G (1983) Transient involvement of signal recognition particle and its receptor in the microsomal membrane prior to protein translocation. Cell 35:677–685

Giraudo CG, Maccioni HJ (2003) Endoplasmic reticulum export of glycosyltransferases depends on interaction of a cytoplasmic dibasic motif with Sar1. Mol Biol Cell 14:3753–3766

Goder V, Spiess M (2001) Topogenesis of membrane proteins: determinants and dynamics. FEBS Lett 504:87–93

Godowski PJ, Leung DW, Meacham LR, Galgani JP, Hellmiss R, Keret R, Rotwein PS, Parks JS, Laron Z, Wood WI (1989) Characterization of the human growth hormone receptor gene and demonstration of a partial gene deletion in two patients with Laron-type dwarfism. Proc Natl Acad Sci USA 86:8083–8087

Goldstein JL, Brown MS, Stone NJ (1977) Genetics of the LDL receptor: evidence that the mutations affecting binding and internalization are allelic. Cell 12:629–641

Goldstein JL, Brown MS (1979) The LDL receptor locus and the genetics of familial hypercholesterolemia. Annu Rev Genet 13:259–89

Görlich D, Prehn S, Hartmann E, Kalies KU, Rapoport TA (1992) A mammalian homolog of SEC61p and SECYp is associated with ribosomes and nascent polypeptides during translocation. Cell 71:489–503

Görlich D, Rapoport TA (1993) Protein translocation into proteoliposomes reconstituted from purified components of the endoplasmic reticulum membrane. Cell 75:615–30

Gorvel JP, Ferrero A, Chambraud L, Rigal A, Bonicel J, Maroux S (1991) Expression of sucrase-isomaltase and dipeptidylpeptidase IV in human small intestine and colon. Gastroenterology 101:618–625

Gow A, Friedrich VL Jr, Lazzarini RA (1994) Many naturally occurring mutations of myelin proteolipid protein impair its intracellular transport. J Neurosci Res 37:574–583

Gow A, Southwood CM, Lazzarini RA (1998) Disrupted proteolipid protein trafficking results in oligodendrocyte apoptosis in an animal model of Pelizaeus-Merzbacher disease. J Cell Biol 140:925–934

Haardt M, Benharouga M, Lechardeur D, Kartner N, Lukacs GL (1999) C-terminal truncations destabilize the cystic fibrosis transmembrane conductance regulator without impairing its biogenesis A novel class of mutation. J Biol Chem 274:21873–21877

Halaban R, Svedine S, Cheng E, Smicum Y, Aron R, Hebert DN (2000) Endoplasmic reticulum retention is a common defect associated with tyrosinase-negative albinism. Proc Natl Acad Sci USA 97:5889–5894

Hammond C, Helenius A (1994a) Folding of VSV G protein: sequential interaction with BiP and calnexin. Science 266:456–458

Hammond C, Helenius A (1994b) Quality control in the secretory pathway: retention of a misfolded viral membrane glycoprotein involves cycling between the ER, intermediate compartment, and Golgi apparatus. J Cell Biol 126:41–52

Harley CA, Tipper DJ (1996) The role of charged residues in determining transmembrane protein insertion orientation in yeast. J Biol Chem 271:24625–24633

Harley CA, Holt JA, Turner R, Tipper DJ (1998) Transmembrane protein insertion orientation in yeast depends on the charge difference across transmembrane segments, their total hydrophobicity, and its distribution. J Biol Chem 273:24963–24971

Hartmann E, Rapoport TA, Lodish HF (1989) Predicting the orientation of eukaryotic membrane-spanning proteins. Proc Natl Acad Sci, USA 86:5786–5790

Hauri HP, Quaroni A, Isselbacher KJ (1979) Biogenesis of intestinal plasma membrane: posttranslational route and cleavage of sucrase-isomaltase. Proc Natl Acad Sci USA 76:5183–5186

Hauri H, Appenzeller C, Kuhn F, Nufer (2000) Lectins and traffic in the secretory pathway. FEBS Lett 476:32–37

Heda GD, Tanwani M, Marino CR (2001) The Delta F508 mutation shortens the biochemical half-life of plasma membrane CFTR in polarized epithelial cells Am J Physiol Cell Physiol. 280:C166–174

Heinrich SU, Mothes W, Brunner J, Rapoport TA (2000) The Sec61p complex mediates the integration of a membrane protein by allowing lipid partitioning of the transmembrane domain. Cell 102:233–244

Heinrich SU, Rapoport TA (2003) Cooperation of transmembrane segments during the integration of a double-spanning protein into the ER membrane. EMBO J 22:3654–3663

Hermann JM, Malkus P, Schekman R (1999) Out of the ER-outfitters, escorts and guides. Trends Cell Biol 9:5–7

Hermosilla R, Schülein R (2001) Sorting functions of the individual cytoplasmic domains of the G protein-coupled vasopressin V2 receptor in Madin Darby canine kidney epithelial cells. Mol Pharmacol 60:1031–1039

Heymann JA, Subramaniam S (1997) Expression, stability, and membrane integration of truncation mutants of bovine rhodopsin. Proc Natl Acad Sci USA 94:4966–48971

Hobbs HH, Russell DW, Brown MS, Goldstein JL (1990) The LDL receptor locus in familial hypercholesterolemia: mutational analysis of a membrane protein. Annu Rev Genet 24:133–170

Horn F, Weare J, Beukers MW, Horsch S, Bairoch A, Chen W, Edvardsen O, Campagne F, Vriend G (1998) GPCRDB: an information system for G protein-coupled receptors. Nucleic Acids Res 26:275–279

Hübner CA, Jentsch TJ (2002) Ion channel diseases Hum Mol Genet 11:2435–2445

Hunziker W, Fumey C (1994) A di-leucine motif mediates endocytosis and basolateral sorting of macrophage IgG Fc receptors in MDCK cells. EMBO J 13:2963–2967

Hull KL, Harvey S (1999) Growth hormone resistance: clinical states and animal models. J Endocrinol 163:165–172

Inglehearn CF, Tarttelin EE, Plant C, Peacock RE, al-Maghtheh M, Vithana E, Bird AC, Bhattacharya SS (1998) A linkage survey of 20 dominant retinitis pigmentosa families: frequencies of the nine known loci and evidence for further heterogeneity. J Med Genet 35:1–5

Innamorati G, Sadeghi H, Birnbaumer M (1996) A fully active nonglycosylated V2 vasopressin receptor. Mol Pharmacol 50:467–473

Jacob R, Zimmer KP, Schmitz J, Naim HY (2000) Congenital sucrase-isomaltase deficiency arising from cleavage and secretion of a mutant form of the enzyme. J Clin Invest 106:281–287

Jensen TJ, Loo MA, Pind S, Williams DB, Goldberg AL, Riordan JR (1995) Multiple proteolytic systems, including the proteasome, contribute to CFTR processing. Cell 83:129–135

Jørgensen MM, Jensen ON, Holst HU, Hansen JJ, Corydon TJ, Bross P, Bolund L, Gregersen N (2000) Grp78 is involved in retention of mutant low density lipoprotein receptor protein in the endoplasmic reticulum. J Biol Chem 275:33861–33868

Kamsteeg EJ, Wormhoudt TA, Rijss JP, van Os CH, Deen PM (1999) An impaired routing of wild-type aquaporin-2 after tetramerization with an aquaporin-2 mutant explains dominant nephrogenic diabetes insipidus. EMBO J 18:2394–2400

Kappeler F, Klopfenstein DR, Foguet M, Paccaud JP, Hauri HP (1997) The recycling of ERGIC-53 in the early secretory pathway ERGIC-53 carries a cytosolic endoplasmic reticulum-exit determinant interacting with COPII. J Biol Chem 272:31801–31808

Kiefer D, Hu X, Dalbey R, Kuhn A (1997) Negatively charged amino acid residues play an active role in orienting the Sec-independent Pf3 coat protein in the *Escherichia coli* inner membrane. EMBO J 16:2197–2204

Klumperman J, Schweizer A, Clausen H, Tang BL, Hong W, Oorschot V, Hauri HP (1998) The recycling pathway of protein ERGIC-53 and dynamics of the ER-Golgi intermediate compartment. J Cell Sci 11:3411–3425

Klumperman J (2000) Transport between ER and Golgi Curr Opin Cell Biol 12:445–449

Knoers NV, Deen PM (2001) Molecular and cellular defects in nephrogenic diabetes insipidus. Pediatr Nephrol 16:1146–1152

Ko YH, Pedersen PL (2001) Cystic fibrosis: a brief look at some highlights of a decade of research focused on elucidating and correcting the molecular basis of the disease. J Bioenerg Biomembr 33:513–521

Köchl R, Alken M, Rutz C, Krause G, Oksche A, Rosenthal W, Schülein R (2002) The signal peptide of the G protein-coupled human endothelin B receptor is necessary for translocation of the N-terminal tail across the endoplasmic reticulum membrane. J Biol Chem 277:16131–16138

Krause G, Hermosilla R, Oksche A, Rutz C, Rosenthal W, Schülein R (2000) Molecular and conformational features of a transport-relevant domain in the C-terminal tail of the vasopressin V2 receptor. Mol Pharmacol 57:232–242

Kuehn MJ, Herrmann JM, Schekman R (1998) COPII-cargo interactions direct protein sorting into ER-derived transport vesicles. Nature 391:187–190

Kuwahara M, Iwai K, Ooeda T, Igarashi T, Ogawa E, Katsushima Y, Shinbo I, Uchida S, Terada Y, Arthus MF, Lonergan M, Fujiwara TM, Bichet DG, Marumo F, Sasaki S (2001) Three families with autosomal dominant nephrogenic diabetes insipidus caused by aquaporin-2 mutations in the C-terminus. Am J Hum Genet 69:738–748

Laron Z, Klinger B, Eshet R, Kaneti H, Karasik A, Silbergeld A (1993) Laron syndrome due to a post-receptor defect: response to IGF-I treatment. Isr J Med Sci 29:757–763

Letts VA, Felix R, Biddlecome GH, Arikkath J, Mahaffey CL, Valenzuela A, Bartlett FS 2nd, Mori Y, Campbell KP, Frankel WN (1998) The mouse stargazer gene encodes a neuronal Ca^{2+}-channel gamma subunit. Nat Genet 19:340–347

Li T, Snyder WK, Olsson JE, Dryja TP (1996) Transgenic mice carrying the dominant rhodopsin mutation P347S: evidence for defective vectorial transport of rhodopsin to the outer segments. Proc Natl Acad Sci USA 93:14176–14181

Loo TW, Ho C, Clarke DM (1995) Expression of a functionally active human renal sodium-calcium exchanger lacking a signal sequence. J Biol Chem 270:19345–19350

Lorenz D, Krylov A, Hahm D, Hagen V, Rosenthal W, Pohl P, Maric K (2003) Cyclic AMP is sufficient for triggering the exocytic recruitment of aquaporin-2 in renal epithelial cells. EMBO Rep 4:88–93

Loffing J, Moyer BD, Reynolds D, Stanton BA (1999) PBA increases CFTR expression but at high doses inhibits Cl^- secretion in Calu-3 airway epithelial cells. Am J Physiol 277:L700–708

Lu Y, Xiong X, Helm A, Kimani K, Bragin A, Skach WR (1998) Co- and posttranslational translocation mechanisms direct cystic fibrosis transmembrane conductance regulator N terminus transmembrane assembly. J Biol Chem 273:568–576

Ma D, Zerangue N, Lin YF, Collins A, Yu M, Jan YN, Jan LY (2001) Role of ER export signals in controlling surface potassium channel numbers. Science 291:316–319

Malkus P, Jiang F, Schekman R (2002) Concentrative sorting of secretory cargo proteins into COPII-coated vesicles. J Cell Biol 159:915–921

Margeta-Mitrovic M, Jan YN, Jan LY (2000) A trafficking checkpoint controls $GABA_B$ receptor heterodimerization. Neuron 27:97–106

Marr N, Bichet DG, Hoefs S, Savelkoul PJ, Konings IB, De Mattia F, Graat MP, Arthus MF, Lonergan M, Fujiwara TM, Knoers NV, Landau D, Balfe WJ, Oksche A, Rosenthal W, Müller D, Van Os CH, Deen PM (2002a) Cell-biologic and functional analyses of five new aquaporin-2 missense mutations that cause recessive nephrogenic diabetes insipidus. J Am Soc Nephrol 13:2267–77

Marr N, Bichet DG, Lonergan M, Arthus MF, Jeck N, Seyberth HW, Rosenthal W, van Os CH, Oksche A, Deen PM (2002b) Heteroligomerization of an aquaporin-2 mutant with wild-type aquaporin-2 and their misrouting to late endosomes/lysosomes explains dominant nephrogenic diabetes insipidus. Hum Mol Genet 11:779–789

Martin PE, Coleman SL, Casalotti SO, Forge A, Evans WH (1999) Properties of connexin26 gap junctional proteins derived from mutations associated with non-syndromal hereditary deafness. Hum Mol Genet 8:2369–76

Martin PE, Mambetisaeva ET, Archer DA, George CH, Evans WH (2000) Analysis of gap junction assembly using mutated connexins detected in Charcot-Marie-Tooth X-linked disease. J Neurochem 74:711–720

Martinez-Ménarguez JA, Geuze HJ, Slot JW, Klumperman J (1999) Vesicular tubular clusters between the ER and Golgi mediate concentration of soluble secretory proteins by exclusion from COPI-coated vesicles. Cell 98:81–90

Martoglio B, Hofmann MW, Brunner J, Dobberstein B (1995) The protein-conducting channel in the membrane of the endoplasmic reticulum is open laterally toward the lipid bilayer. Cell 81:207–214

Matter K, Hunziker W, Mellman I (1992) Basolateral sorting of LDL receptor in MDCK cells: the cytoplasmic domain contains two tyrosine-dependent targeting determinants. Cell 71:741–753

Meacham GC, Lu Z, King S, Sorscher E, Tousson A, Cyr DM (1999) The Hdj-2/Hsc70 chaperone pair facilitates early steps in CFTR biogenesis. EMBO J 18:1492–1505

Meacham GC, Patterson C, Zhang W, Younger JM, Cyr DM (2001) The Hsc70 co-chaperone CHIP targets immature CFTR for proteasomal degradation. Nat Cell Biol 3:100–105

Mehta A, Lu X, Block TM, Blumberg BS, Dwek RA (1997) Hepatitis B virus (HBV) envelope glycoproteins vary drastically in their sensitivity to glycan processing: evidence that alteration of a single N-linked glycosylation site can regulate HBV secretion. Proc Natl Acad Sci USA 94:1822–1827

Merryweather-Clarke AT, Pointon JJ, Shearman JD, Robson KJ (1997) Global prevalence of putative haemochromatosis mutations. J Med Genet 34:275–278

Miller EA, Beilharz TH, Malkus PN, Lee MC, Hamamoto S, Orci L, Schekman R (2003) Multiple cargo binding sites on the COPII subunit Sec24p ensure capture of diverse membrane proteins into transport vesicles. Cell 114:497–509

Molinari M, Helenius A (2000) Chaperone selection during glycoprotein translocation into the endoplasmic reticulum. Science 288:331–333

Moolenaar CE, Ouwendijk J, Wittpoth M, Wisselaar HA, Hauri HP, Ginsel LA, Naim HY, Fransen JA (1997) A mutation in a highly conserved region in brush-border sucrase-isomaltase and lysosomal alpha-glucosidase results in Golgi retention. J Cell Sci 110:557–567

Morello JP, Salahpour A, Laperriere A, Bernier V, Arthus MF, Lonergan M, Petaja-Repo U, Angers S, Morin D, Bichet DG, Bouvier M (2000a) Pharmacological chaperones rescue cell-surface expression and function of misfolded V2 vasopressin receptor mutants. J Clin Invest 105:887–895

Morello JP, Petaja-Repo UE, Bichet DG, Bouvier M (2000b) Pharmacological chaperones: a new twist on receptor folding. Trends Pharmacol Sci 21:466–469

Morello JP, Salahpour A, Petaja-Repo UE, Laperriere A, Lonergan M, Arthus MF, Nabi IR, Bichet DG, Bouvier M (2001) Association of calnexin with wild type and mutant AVPR2 that causes nephrogenic diabetes insipidus. Biochemistry 40:6766–6775

Morello JP, Bichet DG (2001) Nephrogenic diabetes insipidus. Annu Rev Physiol 63:607–30

Moritz OL, Tam BM, Papermaster DS, an Nakayama T (2001) A functional rhodopsin-green fluorescent protein fusion protein localizes correctly in transgenic Xenopus laevis retinal rods and is expressed in a time-dependent pattern. J Biol Chem 276:28242–28251

Mothes W, Heinrich SU, Graf R, Nilsson I, von Heijne G, Brunner J, Rapoport TA (1997) Molecular mechanism of membrane protein integration into the endoplasmic reticulum. Cell 89:523–33

Moussalli M, Pipe SW, Hauri HP, Nichols WC, Ginsburg D, Kaufman RJ (1999) Mannose-dependent endoplasmic reticulum (ER)-Golgi intermediate compartment-53-mediated ER to Golgi trafficking of coagulation factors V and VIII. J Biol Chem 274:32539–32542

Moyer BD, Loffing-Cueni D, Loffing J, Reynolds D, Stanton BA (1999) Butyrate increases apical membrane CFTR but reduces chloride secretion in MDCK cells. Am J Physiol 277:F271–276

Mulders SM, Knoers NV, Van Lieburg AF, Monnens LA, Leumann E, Wuhl E, Schober E, Rijss JP, Van Os CH, Deen PM (1997) New mutations in the AQP2 gene in nephrogenic diabetes insipidus resulting in functional but misrouted water channels. J Am Soc Nephrol 8:242–248

Mulders SM, Bichet DG, Rijss JP, Kamsteeg EJ, Arthus MF, Lonergan M, Fujiwara M, Morgan K, Leijendekker R, van der Sluijs P, van Os CH, Deen PM (1998) An aquaporin-2 water channel mutant which causes autosomal dominant nephrogenic diabetes insipidus is retained in the Golgi complex. J Clin Invest 102:57–66

Muniz M, Nuoffer C, Hauri HP, Riezman H (2000) The Emp24 complex recruits a specific cargo molecule into endoplasmic reticulum-derived vesicles. J Cell Biol 148:925–930

Naef R, Adlkofer K, Lescher B, Suter U (1997) Aberrant protein trafficking in Trembler suggests a disease mechanism for hereditary human peripheral neuropathies. Mol Cell Neurosci 9:13–25

Naef R, Suter U (1999) Impaired intracellular trafficking is a common disease mechanism of PMP22 point mutations in peripheral neuropathies. Neurobiol Dis 6:1–14

Naim HY, Roth J, Sterchi EE, Lentze M, Milla P, Schmitz J, Hauri HP (1988a) Sucrase-isomaltase deficiency in humans Different mutations disrupt intracellular transport, processing, and function of an intestinal brush border enzyme. J Clin Invest 82:667–679

Naim HY, Sterchi EE, Lentze MJ (1988b) Structure, biosynthesis, and glycosylation of human small intestinal maltase-glucoamylase. J Biol Chem 263:19709–19717

Neerman-Arbez M, Johnson KM, Morris MA, McVey JH, Peyvandi F, Nichols WC, Ginsburg D, Rossier C, Antonarakis SE, Tuddenham EG (1999) Molecular analysis of the ERGIC-53 gene in 35 families with combined factor V-factor VIII deficiency. Blood 93:2253–2260

Newton JM, Cohen-Barak O, Hagiwara N, Gardner JM, Davisson MT, King RA, Brilliant M H (2001) Mutations in the human orthologue of the mouse underwhite gene (uw) underlie a new form of oculocutaneous albinism. Am J Hum Gen 69:981–988

Nichols WC, Seligsohn U, Zivelin A, Terry VH, Hertel CE, Wheatley MA, Moussalli MJ, Hauri HP, Ciavarella N, Kaufman RJ, Ginsburg D (1998) Mutations in the ER-Golgi intermediate compartment protein ERGIC-53 cause combined deficiency of coagulation factors V and VIII. Cell 93:61–70

Nichols WC, Terry VH, Wheatley MA, Yang A, Zivelin A, Ciavarella N, Stefanile C, Matsushita T, Saito H, de Bosch NB, Ruiz-Saez A, Torres A, Thompson AR, Feinstein DI, White GC, Negrier C, Vinciguerra C, Aktan M, Kaufman RJ, Ginsburg D, Seligsohn U (1999) Protein ERGIC-53 gene structure and mutation analysis in 19 combined factors V and VIII deficiency families. Blood 93:2261–2266

Nielsen S, Frokiaer J, Marples D, Kwon TH, Agre P, Knepper MA (2002) Aquaporins in the kidney: from molecules to medicine. Physiol Rev 82:205–244

Nilsson I, Witt S, Kiefer H, Mingarro I, von Heijne G (2000) Distant downstream sequence determinants can control N-tail translocation during protein insertion into the endoplasmic reticulum membrane. J Biol Chem 275:6207–6213

Nimpf J, Schneider WJ (2000) From cholesterol transport to signal transduction: low density lipoprotein receptor, very low density lipoprotein receptor, and apolipoprotein E receptor-2. Biochim Biophys Acta 1529:287–298

Nishimura N, Balch WE (1997) A di-acidic signal required for selective export from the endoplasmic reticulum. Science 277:556–558

Nishimura N, Bannykh S, Slabough S, Matteson J, Altschuler Y, Hahn K, Balch WE (1999) A di-acidic (DXE) code directs concentration of cargo during export from the endoplasmic reticulum. J Biol Chem 274:15937–15946

Noiva R (1999) Protein disulfide isomerase: the multifunctional redox chaperone of the endoplasmic reticulum. Semin Cell Dev Biol 10:481–493

Noorwez SM, Kuksa V, Imanishi Y, Zhu L, Filipek S, Palczewski K, Kaushal S (2003) Pharmacological chaperone-mediated in vivo folding and stabilization of the P23H-opsin mutant associated with autosomal dominant retinitis pigmentosa. J Biol Chem 278:14442–14450

Nufer O, Kappeler F, Guldbrandsen S, Hauri HP (2003) ER export of ERGIC-53 is controlled by cooperation of targeting determinants in all three of its domains. J Cell Sci 116:4429–4440

Oetting WS, King RA (1999) Molecular basis of albinism: mutations and polymorphisms of pigmentation genes associated with albinism. Hum Mutat 13:99–115

Okada T, Palczewski K (2001) Crystal structure of rhodopsin: implications for vision and beyond. Curr Opin Struct Biol 11:420–426

Oksche A, Schülein R, Rutz C, Liebenhoff U, Dickson J, Müller H, Birnbaumer M, Rosenthal W (1996) Vasopressin V2 receptor mutants causing X-linked nephrogenic diabetes insipidus: analysis of expression, processing and function. Mol Pharmacol 50:820–828

Oksche A, Dehe M, Schülein R, Wiesner B, Rosenthal W (1998) Folding and cell surface expression of the vasopressin V2 receptor: requirement of the intracellular C-terminus. FEBS Lett 424:57–62

Oksche A, Rosenthal W (1998) The molecular basis of nephrogenic diabetes insipidus. J Mol Med 76:326–327

Oksche A, Leder G, Valet S, Platzer M, Hasse K, Geist G, Krause G, Rosenthal A, Rosenthal W (2002) Variant amino acids in the extracellular loops of murine and human vasopressin V2 receptors account for differences in cell surface expression and ligand affinity. Mol Endocrinol 16:799–813

Olynyk JK, Cullen DJ, Aquilia S, Rossi E, Summerville L, Powell LW (1999) A population-based study of the clinical expression of the hemochromatosis gene. N Engl J Med 341:718–724

Ott CM, Lingappa VR (2002) Integral membrane protein biosynthesis: why topology is hard to predict. J Cell Sci 115:2003–2009

Ouwendijk J, Moolenaar CE, Peters WJ, Hollenberg CP, Ginsel LA, Fransen JA, Naim HY (1996) Congenital sucrase-isomaltase deficiency Identification of a glutamine to proline substitution that leads to a transport block of sucrase-isomaltase in a pre-Golgi compartment. J Clin Invest 97:633–641

Ouzzine M, Magdalou J, Burchell B, Fournel-Gigleux S (1999a) Expression of a functionally active human hepatic UDP-glucuronosyltransferase (UGT1A6) lacking the N-terminal signal sequence in the endoplasmic reticulum. FEBS Lett 454:87–91

Ouzzine M, Magdalou J, Burchell B, Fournel-Gigleux S (1999b) An internal signal sequence mediates the targeting and retention of the human UDP-glucuronosyltransferase 1A6 to the endoplasmic reticulum. J Biol Chem 274:31401–31409

Owens D J, Evans PR (1998) A structural explanation for the recognition of tyrosine based endocytotic signals. Science 282:1327–1332

Pagano A, Rovelli G, Mosbacher J, Lohmann T, Duthey B, Stauffer D, Ristig D, Schuler V, Meigel I, Lampert C, Stein T, Prezeau L, Blahos J. Pin J, Froestl W, Kuhn R, Heid J, Kaupmann K, Bettler B (2001) C-terminal interaction is essential for surface trafficking but not for heteromeric assembly of $GABA_b$ receptors. J Neurosci 21:1189–1202

Palczewski K, Kumasaka T, Hori T, Behnke CA, Motoshima H, Fox BA, Le Trong I, Teller DC, Okada T, Stenkamp RE, Yamamoto M, Miyano M (2000) Crystal structure of rhodopsin: A G protein-coupled receptor. Science 289:739–745

Pan Y, Wilson P, Gitschier J (1994) The effect of eight V2 vasopressin receptor mutations on stimulation of adenylyl cyclase and binding to vasopressin. J Biol Chem 269:31933–31937

Parodi AJ (2000) Protein glucosylation and its role in protein folding. Annu Rev Biochem 69:69–93

Pasyk EA, Foskett JK (1995) Mutant (ΔF508) cystic fibrosis transmembrane conductance regulator Cl- channel is functional when retained in endoplasmic reticulum of mammalian cells. J Biol Chem 270:12347–12350

Paulussen A, Raes A, Matthijs G, Snyders DJ, Cohen N, Aerssens J (2002) A novel mutation (T65P) in the PAS domain of the human potassium channel HERG results in the long QT syndrome by trafficking deficiency. J Biol Chem 277:48610–48616

Peng R, Grabowski R, De Antoni A, Gallwitz D (1999) Specific interaction of the yeast cis-Golgi syntaxin Sed5p and the coat protein complex II component Sec24p of endoplasmic reticulum-derived transport vesicles. Proc Natl Acad Sci USA 96:3751–3756

Petaja-Repo UE, Hogue M, Bhalla S, Laperriere A, Morello JP, Bouvier M (2002) Ligands act as pharmacological chaperones and increase the efficiency of delta opioid receptor maturation. EMBO J 21:1628–37

Phelan JK, Bok D (2000) A brief review of retinitis pigmentosa and the identified retinitis pigmentosa genes Mol Vis 6:116–124

Pind S, Riordan JR, Williams DB (1994) Participation of the endoplasmic reticulum chaperone calnexin (p88,IP90) in the biogenesis of the cystic fibrosis transmembrane conductance regulator. J Biol Chem 269:12784–12788

Postina R, Ufer E, Pfeiffer R, Knoers NV, Fahrenholz F (2000) Misfolded vasopressin V2 receptors caused by extracellular point mutations entail congenital nephrogenic diabetes insipidus. Mol Cell Endocrinol 164:31–39

Riordan JR, Rommens JM, Kerem B, Alon N, Rozmahel R, Grzelczak Z, Zielenski J, Lok S, Plavsic N, Chou JL, Drumm ML, Iannuzzi MC, Collins FS, Tsui LC (1989) Identification of the cystic fibrosis gene: cloning and characterization of complementary DNA. Science 245:1066–1073

Risler JL, Delorme MO, Delacroix H, Henaut A (1988) Amino acid substitutions in structurally related proteins A pattern recognition approach Determination of a new and efficient scoring matrix. J Mol Biol 204:1019–1029

Rodriguez MC, Xie YB, Wang H, Collison K, Segaloff L (1992) Effects of truncations of the cytoplasmic tail of the luteinizing hormone/chorionic gonadotropin receptor on receptor-mediated hormone internalization. Mol Endocrinol 6:327–336

Rosenthal W, Seibold A, Antaramian A, Lonergan M, Arthus MF, Hendy GN, Birnbaumer M, Bichet DG (1992) Molecular identification of the gene responsible for congenital nephrogenic diabetes insipidus. Nature 359:233–235

Rothman JE (1994) Mechanisms of intracellular protein transport. Nature 372:55–63

Roti EC, Myers CD, Ayers RA, Boatman DE, Delfosse SA, Chan EK, Ackerman MJ, January CT, Robertson GA (2002) Interaction with GM130 during HERG ion channel trafficking Disruption by type 2 congenital long QT syndrome mutations Human Ether-a-go-go-Related Gene. J Biol Chem 277:47779–47785

Rubenstein RC, Egan ME, Zeitlin PL (1997) In vitro pharmacologic restoration of CFTR-mediated chloride transport with sodium 4-phenylbutyrate in cystic fibrosis epithelial cells containing ΔF508-CFTR. J Clin Invest 100:2457–2465

Rutz C, Rosenthal W, Schülein R (1999) A single negatively charged residue affects the orientation of a membrane protein in the inner membrane of *Escherichia coli* only when it is located adjacent to a transmembrane domain. J Biol Chem 274:33757–33763

Ryan MC, Notterpek L, Tobler AR, Liu N, Shooter EM (2000) Role of the peripheral myelin protein 22 N-linked glycan in oligomer stability. J Neurochem 75:1465–1474

Sadeghi HM, Innamorati G, Birnbaumer M (1997) An X-linked NDI mutation reveals a requirement for cell surface V_2R expression. Mol Endocrinol 11:706–713

Sadtler S, Laube B, Lashub A, Nicke A, Betz H, Schmalzing G (2003) A basic cluster determines topology of the cytoplasmic M3-M4 loop of the glycine receptor alpha1 subunit. J Biol Chem 278:16782–16790

Sanders CR, Ismail-Beigi F, McEnery MW (2001) Mutations of peripheral myelin protein 22 result in defective trafficking through mechanisms which may be common to diseases involving tetraspan membrane proteins. Biochemistry 40:9453–9459

Sato S, Ward CL, Krouse ME, Wine JJ, Kopito RR (1996) Glycerol reverses the misfolding phenotype of the most common cystic fibrosis mutation. J Biol Chem 271:635–638

Sato M, Hresko R, Mueckler M (1998) Testing the charge difference hypothesis for the assembly of a eucaryotic multispanning membrane protein. J Biol Chem 273:25203–25208

Savage MO, Burren CP, Blair JC, Woods KA, Metherell L, Clark AJ, Camacho-Hubner C (2001) Growth hormone insensitivity: pathophysiology, diagnosis, clinical variation and future perspectives. Horm Res 55 Suppl 2:32–35

Schekman R, Orci L (1996) Coat proteins and vesicle budding. Science 271:1526–1533

Schöneberg T, Liu J, Wess J (1995) Plasma membrane localization and functional rescue of truncated forms of a G protein-coupled receptor. J Biol Chem 270:18000–18006

Schöneberg T, Yun J, Wenkert D, Wess J (1996) Functional rescue of mutant V2 vasopressin receptors causing nephrogenic diabetes insipidus by a co-expressed receptor polypeptide. EMBO J 15:1283–91

Schrag JD, Bergeron JJ, Li Y, Borisova S, Hahn M, Thomas DY, Cygler M (2001) The Structure of calnexin, an ER chaperone involved in quality control of protein folding. Mol Cell 8:633–644

Schülein R, Rutz C, Rosenthal W (1996a) Membrane targeting and determination of transmembrane topology of the human vasopressin V2 receptor. J Biol Chem 271:28844–28852

Schülein R, Liebenhoff U, Müller H, Birnbaumer M, Rosenthal W (1996b) Properties of the human arginine vasopressin V2 receptor after site-directed mutagenesis of its putative palmitoylation site. Biochem J 313:611–616

Schülein R, Hermosilla R, Oksche A, Dehe M, Wiesner B, Krause G, Rosenthal W (1998a) A di-leucine sequence and an upstream glutamate residue in the intracellular C-terminus of the vasopressin V2 receptor are essential for cell surface transport. Mol Pharmacol 54:525–535

Schülein R, Lorenz D, Oksche A, Wiesner B, Hermosilla R, Ebert J, Rosenthal W (1998b) Polarized cell surface expression of the G protein-coupled vasopressin V2 receptor in Madin Darby canine kidney cells. FEBS Lett 441:170–176

Schülein R, Zühlke K, Oksche A, Furkert J, Rosenthal W (2000) The role of conserved extracellular cysteine residues in vasopressin V2 receptor function and properties of two naturally occurring mutant receptors with additional extracellular cysteine residues. FEBS Lett 466:101–106

Schülein R, Zühlke K, Krause G, Rosenthal W (2001) Functional rescue of the nephrogenic diabetes insipidus causing vasopressin V2 receptor mutants G185C and R202C by a second site suppressor mutation. J Biol Chem 276:8384–8392

Schwiebert EM, Benos DJ, Egan ME, Stutts MJ, Guggino WB (1999) CFTR is a conductance regulator as well as a chloride channel. Physiol Rev 79:S145–S66

Scott DB, Blanpied TA, Swanson GT, Zhang C, Ehlers MD (2001) An NMDA receptor ER retention signal regulated by phosphorylation and alternative splicing. J Neurosci 21:3063–72

Sevier CS, Weisz OA, Davis M, Machamer CE (2000) Efficient export of the vesicular stomatitis virus G protein from the endoplasmic reticulum requires a signal in the cytoplasmic tail that includes both tyrosine-based and di-acidic motifs. Mol Biol Cell 11:13–22

Sharma M, Benharouga M, Hu W, Lukacs GL (2001) Conformational and temperature-sensitive stability defects of the ΔF508 cystic fibrosis transmembrane conductance regulator in post-endoplasmic reticulum compartments. J Biol Chem 276:8942–8950

Sheppard DN, Welsh MJ (1999) Structure and function of the CFTR chloride channel. Physiol Rev 79:S23–S45

Simon DB, Lu Y, Choate KA, Velazquez H, Al-Sabban E, Praga M, Casari G, Bettinelli A, Colussi G, Rodriguez-Soriano J, McCredie D, Milford D, Sanjad S, Lifton RP (1999) Paracellin-1, a renal tight junction protein required for paracellular Mg^{2+} resorption. Science 285:103–106

Smilde TJ, van Wissen S, Wollersheim H, Kastelein JJ, Stalenhoef AF (2001) Genetic and metabolic factors predicting risk of cardiovascular disease in familial hypercholesterolemia. Neth J Med 59:184–195

Sousa MC, Ferrero-Garcia MA, Parodi AJ (1992) Recognition of the oligosaccharide and protein moieties of glycoproteins by the UDP-Glc:glycoprotein glucosyltransferase. Biochemistry 31:97–105

Sousa M, Parodi AJ (1995) The molecular basis for the recognition of misfolded glycoproteins by the UDP-Glc:glycoprotein glucosyltransferase. EMBO J 14:4196–4203

Southwood C, Gow A (2001) Molecular pathways of oligodendrocyte apoptosis revealed by mutations in the proteolipid protein gene. Microsc Res Tech 52:700–708

Spiess M (1995) Heads or tails—what determines the orientation of proteins in the membrane FEBS Lett 369:76–79

Splawski I, Tristani-Firouzi M, Lehmann MH, Sanguinetti MC, Keating MT (1997) Mutations in the hminK gene cause long QT syndrome and suppress I_{Ks} function. Nat Genet 17:338–340

Splawski I, Shen J, Timothy KW, Lehmann MH, Priori S, Robinson JL, Moss AJ, Schwartz PJ, Towbin JA, Vincent GM, Keating MT (2000) Spectrum of mutations in long-QT syndrome genes KVLQT1, HERG, SCN5A, KCNE1, and KCNE2. Circulation 102:1178–1185

Spodsberg N, Jacob R, Alfalah M, Zimmer KP, Naim HY (2001) Molecular basis of aberrant apical protein transport in an intestinal enzyme disorder. J Biol Chem 276:23506–23510

Springer S, Schekman R (1998) Nucleation of COPII vesicular coat complex by endoplasmic reticulum to Golgi vesicle SNAREs. Science 281:698–700

Springer S, Spang A, Schekman R (1999) A primer on vesicle budding. Cell 97:145–148

Spritz RA, Strunk KM, Giebel LB, King RA (1990) Detection of mutations in the tyrosinase gene in a patient with type IA oculocutaneous albinism. N Engl J Med 332:1724–1728

Stockklausner C, Ludwig J, Ruppersberg JP, Klocker N (2001) A sequence motif responsible for ER export and surface expression of Kir20 inward rectifier K^+ channels. FEBS Lett 493:129–133

Sung CH, Makino C, Baylor D, Nathans J (1994) A rhodopsin gene mutation responsible for autosomal dominant retinitis pigmentosa results in a protein that is defective in localization to the photoreceptor outer segment. J Neurosci 14:5818–5833

Tam BM, Moritz OL, Hurd LB, Papermaster DS (2000) Identification of an outer segment targeting signal in the COOH terminus of rhodopsin using transgenic Xenopus laevis. J Cell Biol 151:1369–1380

Tamarappoo BK, Verkman AS (1998) Defective aquaporin-2 trafficking in nephrogenic diabetes insipidus and correction by chemical chaperones. J Clin Invest 15:2257–2267

Tamarappoo BK, Yang B, Verkman AS (1999) Misfolding of mutant aquaporin-2 water channels in nephrogenic diabetes insipidus. J Biol Chem 274:34825–34831

Thomson CE, Montague P, Jung M, Nave KA, Griffiths IR (1997) Phenotypic severity of murine Plp mutants reflects in vivo and in vitro variations in transport of PLP isoproteins. Glia 20:322–332

Tobler AR, Notterpek L, Naef R, Taylor V, Suter U, Shooter EM (1999) Transport of Trembler-J mutant peripheral myelin protein 22 is blocked in the intermediate compartment and affects the transport of the wild-type protein by direct interaction. J Neurosci 19:2027–2036

Towbin JA, Wang Z, Li H (2001) Genotype and severity of long QT syndrome. Drug Metab Dispos 29, 574–579

Toyofuku K, Wada I, Valencia JC, Kushimoto T, Ferrans VJ, Hearing VJ (2001a) Oculocutaneous albinism types 1 and 3 are ER retention diseases: mutation of tyrosinase or Tyrp1 can affect the processing of both mutant and wild-type proteins. FASEB J 15:2149–2161

Toyofuku K, Wada I, Spritz RA, Hearing VJ (2001b) The molecular basis of oculocutaneous albinism type 1 (OCA1): sorting failure and degradation of mutant tyrosinases results in a lack of pigmentation. Biochem J 355:259–269

Treem WR (1995) Congenital sucrase-isomaltase deficiency. J Pediatr Gastroenterol Nutr 21:1–14

Trombetta SE, Parodi AJ (1992) Purification to apparent homogeneity and partial characterization of rat liver UDP-glucose:glycoprotein glucosyltransferase. J Biol Chem 267:9236–9240

Trombetta ES, Helenius A (1998) Lectins as chaperones in glycoprotein folding. Curr Opin Struct Biol 8:587–592

Tu L, Wang J, Helm A, Skach WR, Deutsch C (2000) Transmembrane biogenesis of Kv13. Biochemistry 39:824–36

Unson CG, Cypess AM, Kim HN, Goldsmith PK, Carruthers CJ, Merrifield RB, Sakmar TP (1995) Characterization of deletion and truncation mutants of the rat glucagon receptor Seven transmembrane segments are necessary for receptor transport to the plasma membrane and glucagon binding. J Biol Chem 270:27720–27727

van den Berg B, Clemons WM Jr, Collinson I, Modis Y, Hartmann E, Harrison SC, Rapoport TA (2004) X-ray structure of a protein-conducting channel Nature 427:36–44

van den Ouweland AM, Dreesen JC, Verdijk M, Knoers NV, Monnens LA, Rocchi M, van Oost BA (1992) Mutations in the vasopressin type 2 receptor gene (AVPR2) associated with nephrogenic diabetes insipidus. Nat Genet 2:99–102

van Klompenburg W, Nilsson I, von Heijne G, de Kruijff B (1997) Anionic phospholipids are determinants of membrane protein topology. EMBO J 16:4261–4266

van Leeuwen JE, Kearse KP (1996) The related molecular chaperones calnexin and calreticulin differentially associate with nascent T cell antigen receptor proteins within the endoplasmic reticulum. J Biol Chem 271:25345–25349

van Soest S, Westerveld A, de Jong PT, Bleeker-Wagemakers EM, Bergen AA (1999) Retinitis pigmentosa: defined from a molecular point of view. Surv Ophthalmol 43:321–334

Verkman AS, Mitra AK (2000) Structure and function of aquaporin water channels. Am J Physiol Renal Physiol 278:F13–28

von Heijne G, Gavel Y (1988) Topogenic signals in integral membrane proteins Eur J Biochem 174:671–678

von Heijne G (1994) Membrane proteins: from sequence to structure. Annu Rev Biophys Biomol Struct 23:167–192

Votsmeier C, Gallwitz D (2001) An acidic sequence of a putative yeast Golgi membrane protein binds COPII and facilitates ER export. EMBO J 20:6742–50

Waheed A, Parkkila S, Saarnio J, Fleming RE, Zhou XY, Tomatsu S, Britton RS, Bacon BR, Sly WS (1999) Association of HFE protein with transferrin receptor in crypt enterocytes of human duodenum. Proc Natl Acad Sci USA 96, 1579–1584

Waheed A, Parkkila S, Zhou XY, Tomatsu S, Tsuchihashi Z, Feder JN, Schatzman RC, Britton RS, Bacon BR, Sly WS (1997) Hereditary hemochromatosis: effects of C282Y and H63D mutations on association with β_2-microglobulin, intracellular processing, and cell surface expression of the HFE protein in COS-7 cells. Proc Natl Acad Sci USA 94:12384–12389

Wallin E, von Heijne G (1995) Properties of N-terminal tails in G-protein coupled receptors: a statistical study. Protein Eng 8:693–698

Walter P, Blobel G (1981) Translocation of proteins across the endoplasmic reticulum III Signal recognition protein (SRP) causes signal sequence-dependent and site-specific arrest of chain elongation that is released by microsomal membranes. J Cell Biol 91:557–61

Wang Q, Curran ME, Splawski I, Burn TC, Millholland JM, VanRaay TJ, Shen J, Timothy KW, Vincent GM, de Jager T, Schwartz PJ, Toubin JA, Moss AJ, Atkinson DL, Landes GM, Connors TD, Keating MT (1996) Positional cloning of a novel potassium channel gene: KVLQT1 mutations cause cardiac arrhythmias. Nat Genet 12:17–23

Wessels HP, Spiess M (1988) Insertion of a multispanning membrane protein occurs sequentially and requires only one signal sequence. Cell 55:61–70

White TW (2000) Functional analysis of human Cx26 mutations associated with deafness. Brain Res Rev 32:181–183

Wieland FT, Gleason ML, Serafini TA, Rothman JE (1987) The rate of bulk flow from the endoplasmic reticulum to the cell surface. Cell 50:289–300

Wojcik J, Berg MA, Esposito N, Geffner ME, Sakati N, Reiter EO, Dower S, Francke U, Postel-Vinay MC, Finidori J (1998) Four contiguous amino acid substitutions, identified in patients with Laron syndrome, differently affect the binding affinity and intracellular trafficking of the growth hormone receptor. J Clin Endocrinol Metab 83:4481–4489

Woods KA, Dastot F, Preece MA, Clark AJ, Postel-Vinay MC, Chatelain PG, Ranke MB, Rosenfeld RG, Amselem S, Savage MO (1997) Phenotype:genotype relationships in growth hormone insensitivity syndrome. J Clin Endocrinol Metab 82:3529–3535

Yamamoto K, Fujii R, Toyofuku Y, Saito T, Koseki H, Hsu VW, Aoe T (2001) The KDEL receptor mediates a retrieval mechanism that contributes to quality control at the endoplasmic reticulum. EMBO J 20:3082–3091

Yamashita F, Horie M, Kubota T, Yoshida H, Yumoto Y, Kobori A, Ninomiya T, Kono Y, Haruna T, Tsuji K, Washizuka T, Takano M, Otani H, Sasayama S, Aizawa Y (2001) Characterization and subcellular localization of KCNQ1 with a heterozygous mutation in the C terminus. J Mol Cell Cardiol 33:197–207

Yang Y, Janich S, Cohn JA, Wilson JM (1993) The common variant of cystic fibrosis transmembrane conductance regulator is recognized by hsp70 and degraded in a pre-Golgi nonlysosomal compartment. Proc Nat Acad Sci 90:9480–9484

Yeaman C, Grindstaff KK, Nelson J (1999) New perspectives on mechanisms involved in generating epithelial polarity. Physiol Rev 79:73–97

Yokoyama K, Yamauchi A, Izumi M, Itoh T, Ando A, Imai E, Kamada T, Ueda N (1996) A low-affinity vasopressin V2-receptor gene in a kindred with X-linked nephrogenic diabetes insipidus. J Am Soc Nephrol 7:410–414

Young P, Suter U (2001) Disease mechanisms and potential therapeutic strategies in Charcot-Marie-Tooth disease. Brain Res Brain Res Rev 36:213–221

Yuan H, Michelsen K, Schwappach B (2003) 14-3-3 dimers probe the assembly status of multimeric membrane proteins. Curr Biol 13:638–646

Zeitlin PL (2000) Future pharmacological treatment of cystic fibrosis. Respiration 67:351–357

Zerangue N, Schwappach B, Jan YN, Jan LY (1999) A new ER trafficking signal regulates the subunit stoichiometry of plasma membrane K_{ATP} channels. Neuron 22:537–48

Zielenski J (2000) Genotype and phenotype in cystic fibrosis. Respiration 67:117–133

Zhang JX, Braakman I, Matlack KE, Helenius A (1997) Quality control in the secretory pathway: the role of calreticulin, calnexin and BiP in the retention of glycoproteins with C-terminal truncations. Mol Biol Cell 8:1943–1954

Zhou Z, Gong Q, Epstein ML, January CT (1998) HERG channel dysfunction in human long QT syndrome Intracellular transport and functional defects. J Biol Chem 273:21061–21066

Zhou Z, Gong Q, January CT (1999) Correction of defective protein trafficking of a mutant HERG potassium channel in human long QT syndrome Pharmacological and temperature effects. J Biol Chem 274:31123–31126

Zuccon L, Corsi B, Levi S, Mattioli M, Fracanzani AL, Corti A, Albertini A, Sampietro M, Fargion S, Arosio P (2000) Immunohistochemistry of HFE in the duodenum of C282Y homozygotes with antisera for recombinant HFE protein. Haematologica 85:346–351

Rev Physiol Biochem Pharmacol (2004) 151:93–107
DOI 10.1007/s10254-004-0023-7

L. Schild

The epithelial sodium channel: from molecule to disease

Published online: 14 May 2004

Abstract Genetic analysis has demonstrated that Na absorption in the aldosterone-sensitive distal nephron (ASDN) critically determines extracellular blood volume and blood pressure variations. The epithelial sodium channel (ENaC) represents the main transport pathway for Na^+ absorption in the ASDN, in particular in the connecting tubule (CNT), which shows the highest capacity for ENaC-mediated Na^+ absorption. Gain-of-function mutations of ENaC causing hypertension target an intracellular proline-rich sequence involved in the control of ENaC activity at the cell surface. In animal models, these ENaC mutations exacerbate Na^+ transport in response to aldosterone, an effect that likely plays an important role in the development of volume expansion and hypertension. Recent studies of the functional consequences of mutations in genes controlling Na^+ absorption in the ASDN provide a new understanding of the molecular and cellular mechanisms underlying the pathogenesis of salt-sensitive hypertension.

Introduction

According to the classical cell model of Na^+ transport in tight epithelia (Koefoed-Johnson 1958) originally described by Koefoed-Johnson and Ussing, Na^+ ions enter the cell across the apical membrane by diffusion through a Na^+-selective apical conductance. Measurements of current noise in Na^+ transporting tight epithelia provided the first evidence that the apical Na^+-selective conductance involves the presence of active Na^+ channels (Palmer et al. 1980). These epithelial Na^+ channels (ENaC) were then characterized at the single channel level in the apical membrane of rat cortical collecting duct (CCD) principal cells, using the patch clamp technique. The ENaC was found to have a low single channel conductance of 5 pS, a high Na^+ over K^+ selectivity, and a slow gating with lifetimes of the open and closed states lasting several seconds (Palmer and Frindt 1986).

L. Schild (✉)
Department of Pharmacology and Toxicology, University of Lausanne, Rue du Bugnon 27, 1005 Lausanne, Switzerland
e-mail: Laurent.Schild@ipharm.unil.ch · Tel.: +41-21-6925380 · Fax: +41-21-6925355

The primary structure of ENaC was identified, using a distal colon cDNA library and the *Xenopus* oocyte expression system (Canessa et al. 1993; Lingueglia et al. 1993). After the cloning of the α subunit of the ENaC, subsequent functional complementation experiments were necessary to identify β and γ subunits that are required for the maximal expression of the ENaC channel (Canessa et al. 1994). Quite remarkably, the expression of α, β, and γ ENaC subunits in *Xenopus* oocytes reconstituted all the biophysical properties (low conductance, high Na^+ selectivity) and the pharmacological sensitivity to amiloride of the epithelial sodium channel present in the CCD.

Sequence comparison between the ENaC subunits was consistent with an homologous primary structure with ~30% identity at the amino acid level. ENaC belongs to a larger family of ion channels, the ENaC/DEG gene family that is found in metazoa only. This family of ion channels shows a wide tissue distribution and an apparent high functional heterogeneity among its different members (Kellenberger and Schild 2002).

Analysis of the membrane topology of ENaC shows two transmembrane α helices, TM1 and TM2, leaving the amino- and the carboxy-termini in the cytoplasm. A characteristic feature of the members of the ENaC/DEG gene family is the long amino acid sequence separating the two transmembrane helices TM1 and TM2, which represent more than half of the mass of the protein. This stretch of amino acids is believed to face the luminal side of the epithelial kidney cell, but presently no structural information is available that firmly excludes the possibility that a significant portion of this large loop could be embedded in the membrane.

The ENaC shares similarities with other cation channels like K^+ channels: they are heteromultimeric channels made of subunits with two transmembrane spanning segments. Strong evidence supports a heterotetrameric architecture for ENaC with 2α, 1β, and 1γ subunits, arranged pseudosymmetrically around the channel pore (Firsov et al. 1998; Kosari et al. 1998; Coscoy et al. 1998; Dijkink et al. 2002).

Physiology

ENaC in the kidney

Under normal conditions of hydration and salt intake, the kidney reabsorbs 95%–98% of the filtered load of Na^+, the proximal tubule and the thick ascending limb accounting for more than 70% of the reabsorption of Na^+ ions. The remaining load of filtered Na^+ is reabsorbed in the distal convoluted tubule (DCT) and in the aldosterone-sensitive distal nephron (ASDN), which includes the late distal convoluted tubule (late DCT), the connecting tubule (CNT), and the CCD . Under these normal conditions, Na^+ absorption in the ASDN represents only a low percentage of the total filtered load. However, when salt and water intake is reduced, Na^+ absorption in the ASDN can dramatically increase, in order to balance renal salt excretion with the daily intake. This fine tuning of Na^+ handling by the ASDN is mainly under the control of aldosterone and vasopressin (Verrey et al. 2001; Hawk et al. 1996). Thus, despite the relatively low fraction of the filtered Na^+ reabsorbed in the ASDN under physiological conditions, the ability of the distal nephron to retain sodium in response to a drop in the daily intake, points to a critical role of this nephron segment in the maintenance of salt and water homeostasis.

Specific antibodies raised against ENaC subunits clearly recognize proteins in the DCT, CNT, and CCD (Loffing et al. 2000a, 2001a; Loffing and Kaissling 2003; Hager et al. 2001). Expression of ENaC is most predominant in the late DCT and CNT, then progressively decreases along the rat CCD, suggesting an axial heterogeneity in ENaC expression along the ASDN (Duc et al. 1994). This is supported by the observation that Na^+ reabsorption is 3–4 times higher in the rabbit CNT than in the CCD (Almeida and Burg 1982). Recently, ENaC activity was investigated in principal cells of rat CNT, and was found to be also several times higher than in the CCD of the same animals, suggesting that the Na^+ absorptive capacity of the CNT is up to 10 times higher than that of the CCD (Frindt and Palmer 2003). Finally, principal cells of the inner medullary collecting duct (IMCD) show significant immunolabeling for the three α, β, and γ ENaC subunits, with an apparent decrease in the labeling along the axis of the IMCD (Hager et al. 2001).

ENaC gene targeting in mice with complete knockout or decreased expression of α, β, γ subunits further demonstrated the important role that ENaC plays in the maintenance of salt homeostasis (Hummler et al. 1996; Pradervand et al. 1999; McDonald et al. 1999; Barker et al. 1998). The α, β, or γ ENaC knockout mice show a marked renal loss of Na^+ despite elevated plasma levels of aldosterone and hyperkalemia. These mouse models confirm the importance of each of the three α, β, and γ ENaC subunits in mediating Na^+ absorption and K^+ secretion in the distal nephron under the control of aldosterone.

Recently, a tissue-specific knockout of the α ENaC subunit in the CCD has been generated in the mouse (Rubera et al. 2003). The absence of α ENaC protein in the CCD and OMCD, together with an undetectable ENaC activity in these nephron segments, has surprisingly little effect on Na^+ homeostasis. These CCD-specific knockout mice remain in Na^+ and K^+ balance without significant renal loss of Na^+ ions, even under dietary salt restriction. These observations suggest that a loss of function of ENaC in the collecting duct can be compensated by the ENaC-mediated Na^+ absorption in the upstream part of the ASDN, namely the DCT and the CNT. This is consistent with the idea that the CNT has a high capacity for ENaC-mediated Na^+ absorption, which can compensate for Na^+ transport defects in more distal parts of the ASDN. Unfortunately, no tissue-specific knockout of ENaC targeting the CNT is yet available to test this hypothesis.

Active ENaC at the apical membrane of the ASDN epithelial cells allows the electrogenic movement of Na^+ ions from the tubule lumen into the cell. This electrogenic, amiloride-sensitive transport of Na^+ generates a lumen-negative transepithelial potential, as recorded in the isolated and perfused CCDs. This lumen-negative potential represents a favorable electrochemical driving force across the apical membrane for secretion of K^+ ions from the cell into the lumen through the K^+ channel ROMK1. The coupling between ENaC-mediated Na^+ absorption and K^+ secretion in the ASDN is shown in ENaC knockout mice as a decrease in urinary K^+ excretion and a severe hyperkalemia.

Subcellular localization

Under conditions of normal to high salt intake (low plasma aldosterone), ENaC-mediated Na^+ absorption in the ASDN can be measured, but is likely to be quite low: an amiloride-sensitive Na^+ absorption is detectable in the late distal tubule of the rat and in the CCD of rabbits (Grantham et al. 1970; Costanzo 1984). However, no ENaC activity was measured in principal cells of rat CNT and CCD, using the patch clamp technique under similar conditions (Frindt et al. 2002; Frindt and Palmer 2003). Immunolabeling of ENaC revealed that under these conditions, β and γ ENaC subunits are distributed throughout the cytoplasm, whereas α ENaC labeling is restricted to the apical part of the principal cell, likely including the plasma membrane (Hager et al. 2001). Other studies have failed to detect α ENaC in the apical membrane of ASDN by immunostaining or electron microscopy (Hager et al. 2001; Loffing et al. 2000b). These studies indicate that under conditions of normal-to-high salt intake, the density of active ENaC channels at the apical membrane of principal cells is low, at the limit of the detection level by standard immunolabeling, or patch clamp techniques.

A dietary salt restriction raises plasma aldosterone more than twofold. Aldosterone induces the synthesis of α ENaC and promotes the translocation of all three ENaC subunits to the apical membrane of the ASDN. The appearance of newly synthesized ENaC at the apical membrane is evidenced by recordings of ENaC currents by patch clamp in rat CCD principal cells (Frindt et al. 2001, 2002). The half-life of ENaC at the cell surface is approximately 1 h (Hanwell et al. 2002; May et al. 1997).

The first step of the aldosterone action leading to the recruitment of active ENaC at the cell surface is the binding of the hormone to the cytosolic mineralocorticoid receptor (MR receptor). The hormone-receptor complex is then translocated to the nucleus where transcriptional events occur, including an increase in ENaC mRNAs, together with other aldosterone-induced transcripts (AITs), or alternatively a decrease in expression of transcripts, i.e., the aldosterone-repressed transcripts (ARTs; Robert-Nicoud et al. 2001). The AITs and ARTs are key elements for the ENaC activation at the apical membrane and the up-regulation of the Na^+-K^+/ATPase at the basolateral membrane underlying the stimulation of a transepithelial Na^+ flux (Verrey et al. 2001). Aldosterone rapidly increases the synthesis in principal cells of the mouse CCD of a serine/threonine kinase named "serum- and glucocorticoid-regulated kinase" (SGK) (Chen et al. 1999; Naray et al. 1999). SGK was demonstrated to stimulate ENaC when coexpressed in *Xenopus* oocytes (Alvarez de la Rosa et al. 1999). The aldosterone-induced increase in SGK nicely correlates with the marked redistribution of ENaC from a cytoplasmic pool to the apical membrane (Loffing et al. 2001b). This increase in ENaC channel density at the cell surface is thought to contribute essentially to the early response to aldosterone, which increases Na^+ transport within 1–3 h of stimulation by the hormone.

Vasopressin increases both water and Na^+ absorption in the distal nephron (Hawk et al. 1996; Nielsen et al. 1995). Binding of vasopressin to vasopressin receptors (V2 receptors) at the basolateral membrane of principal cells causes a redistribution of aquaporins (AQP-2) from an intracellular compartment to the apical membrane, leading to an increase in cell membrane water permeability. In parallel, the stimulation of the V2 receptors increases ENaC activity at the apical membrane, which accounts for the enhanced Na^+ absorption (Morris and Schafer 2002). Both vasopressin effects on AQP-2 and ENaC are mediated by cAMP and can be mimicked by cAMP analogues or forskolin (Auberson et al. 2003; van Balkom et al. 2004). Beside this signaling pathway, vasopressin also induces the expression of a gene network that might be involved in the long-term regulation of Na^+ absorption (Robert-Nicoud et al. 2001). Expression of genes such as VIP32 or RAMP3 is enhanced, but their precise role in the modulation of ENaC activity remains to be established.

ENaC in other organs

The airway epithelia absorb Na^+ ions via an amiloride-sensitive electrogenic transport. In adult rats and humans the α, β, and γ ENaC subunits are highly expressed in small and medium-size airways (Farman et al. 1997; Talbot et al. 1999; Burch et al. 1995). Apical

Na^+ transport by ENaC is important for the maintenance of the composition and the volume of the airway surface liquid (ASL), which forms a thin layer of liquid that coats the airways' epithelium (Boucher 2004).

At birth, the amiloride-sensitive electrogenic Na^+ transport is important to clear the liquid that fills the alveoli and the airways of the fetal lung. Messenger RNAs for α, β, and γ ENaC can be detected in the mouse fetal lung, at around days 15–17 of gestation and expression of ENaC subunits (mainly α and γ ENaC) sharply increases in the late fetal and early postnatal life when the lung turns from a secretory to an absorptive organ (Talbot et al. 1999). The physiological role of ENaC in lung liquid balance was clearly demonstrated in mice in which the α ENaC gene was inactivated by homologous recombination (Hummler et al. 1996). These α ENaC knock-out mice die soon after birth from a respiratory failure, due to a severe defect in the clearance of the fetal liquid that fills the lungs, associated with severe plasma electrolyte imbalance. These studies suggest that at birth α ENaC in the mouse fetal lung is important for Na^+ absorption. The disruption of the β and γ ENaC gene loci results in a slower clearance of the fetal lung liquid at birth but does not severely affect the blood gas parameters. The β or γ ENaC knock-out mice die slightly later than the α knockout, from severe electrolyte imbalance, namely hyperkalemia due to deficient renal K^+ secretion (Barker et al. 1998; McDonald et al. 1999). Thus, in contrast to the kidney, the Na^+ transport in the lung can be maintained efficiently by only two functional α-β or α-γ ENaC genes.

The colon is a tight epithelium absorbing Na^+ by an electrogenic transport, sensitive to amiloride and stimulated by aldosterone (Epple et al. 2000). Initially the α, β, and γ ENaC cDNAs were isolated from a rat distal colon cDNA library because of the high level of channel expression in this tissue, suggesting that the distal colon is an important site for Na^+ reabsorption. The contribution of the distal colon to the maintenance of the Na^+ homeostasis is not clear, and differences may exist between species.

In salivary glands, α, β, and γ ENaCs are detected at the RNA and protein levels in the apical membrane of the striated and interlobular ducts (Duc et al. 1994). An amiloride-sensitive Na^+ conductance has been described in mandibular duct cells, using the whole-cell patch-clamp technique (Dinudom et al. 1993).

Salt taste is transduced by direct amiloride-sensitive influx of Na^+ in the taste cells of the fungiform papillae of the anterior part of the tongue, suggesting the presence of an amiloride-sensitive Na^+ channel (Lindemann 1996). The three α, β, and γ ENaC subunits are expressed in the taste receptor cells of the fungiform papillae. Finally, the three ENaC subunits are expressed in mammalian keratinocytes, as well as in epithelial cells of hair follicles and sweat glands. The role of ENaC in the skin is still unclear, as well as the role of ENaC in the inner ear, where it is detected in the epithelial cells lining the scala media of the cochlea (for review, see Kellenberger and Schild 2002).

Pathophysiology

Renal salt transport and hypertension

Blood pressure is controlled by multiple mechanisms. The autonomous nervous system protects against major rapid changes in blood pressure, whereas long-term regulation of blood pressure within hours or days is achieved by the renin-angiotensin aldosterone system and the kidneys (Guyton 1992). The kidneys play a major role in blood pressure regulation: they control the extracellular fluid volume by balancing the urinary Na^+ and water excretion with the daily intake of fluid and salt. This precise balance is obtained for a single blood pressure value. Normally, when blood pressure raises, the kidneys excrete more Na^+ and fluid than the amount ingested. Consequently, the extracellular fluid volume decreases, as well as the cardiac output, and blood pressure drops. Hypertension starts to develop when higher blood pressure is required to maintain a balance between ingestion and excretion of Na^+ and fluid by the kidneys.

The precise renal mechanisms involved in the fine control of Na^+ balance still remain to be identified at the cellular and molecular levels. During the last decade, genetic studies of mendelian syndromes associating kidney diseases and defective blood pressure controls have identified genes involved in Na^+ absorption in the distal nephron causing hypo- or hypertension (Lifton 1996). Remarkably, these genes, whose mutations cause rare mendelian forms of hypertension, encode key enzymes, receptors, or channels in the renin-angiotensin-aldosterone cascade and in its target kidney epithelium, the ASDN. The hereditary forms of hypertension include the glucocorticoid-remediable aldosteronism (GRA), the apparent mineralocorticoid excess (AME), and pseudoaldosteronism (or Liddle syndrome; for review, see Lifton et al. 2001).

GRA patients have increased plasma aldosterone levels and a severe hypertension. The hyperaldosteronism is due to gene duplication of the aldosterone synthesis and 11β-hydroxydase genes, generating a novel gene that places the synthesis of aldosterone under the control of corticotrophin. The apparent AME is an autosomal recessive disorder characterized by a moderate to severe hypertension with very low plasma aldosterone levels. In case of AME, the elevated blood pressure can be normalized by the aldosterone antagonist spironolactone. This form of mendelian hypertension is associated with mutations in the 11β-hydroxysteroid-dehydrogenase, a key enzyme present in aldosterone-sensitive kidney cells to protect stimulation of the mineralocorticoid receptor (MR) by cortisol.

Another inherited form of low renin, low aldosterone hypertension exacerbated by pregnancy is due to mutations in the ligand binding domain of the mineralocorticoid receptor, leading to receptor activation by hormones other than aldosterone, such as progesterone, for instance (Geller et al. 2000). Finally, pseudoaldosteronism, described by G. Liddle in the early sixties (Liddle syndrome), associates low renin–low aldosterone hypertension, hypokalemia, and metabolic alkalosis. This syndrome was attributed to an abnormally high Na^+ absorption in the distal nephron. Mutations causing Liddle syndrome are found in the SCNN1B and SCNN1G genes encoding the β and γ ENaC subunits (Shimkets et al. 1994).

The identification of the genetic basis of these rare forms of hypertension further stresses the importance of the ASDN in the maintenance of the Na^+ balance and the long term control of blood pressure.

ENaC and salt-losing nephropathy

The type 1 pseudohypoaldosteronism (PHA-1) is characterized in the first week of life by severe dehydration, hyponatremia, and hyperkalemia resistant to mineralocorticoids. Two clinically distinct forms of PHA-1 have been described, an autosomal recessive form affecting multiple organs (systemic form), including kidneys, colon, salivary glands, and sweat ducts, and an autosomal dominant form, characterized by salt loss in the kidneys only (Hanukoglu 1991). Patients with a systemic form of PHA-1 also have frequent lower respiratory tract infections, but so far no hearing loss or skin alterations have been clinically observed in these patients, as it could be expected from ENaC tissue expression. The systemic form is caused by homozygous mutations in α, β, and γ ENaC subunits (Chang et al. 1996). The autosomal dominant form of PHA-1 is caused by nonsense or frameshift mutations of the mineralocorticoid receptors and is consistent with salt loss restricted to the kidneys (Geller et al. 1998).

Molecular basis of "*ENaCopathies*"

In addition to the clear demonstration from genetic analysis of the Liddle syndrome and PHA-1 of the physiological role of ENaC in the maintenance of Na^+ homeostasis, the identification of mutations associated with ENaC dysfunction also provided new information regarding the molecular and cellular basis of Na^+ and K^+ transport in the ASDN.

ENaC gain-of-function mutations

The ENaC mutations associated with the Liddle syndrome include premature stop codons, and frameshift or missense mutations that delete or substitute residues in a conserved PPP.Y sequence in the cytosolic C-terminus of β and γ ENaC subunits called PY motif (Rossier et al. 2001). A remarkable observation was that the expression of ENaC carrying the mutations causing the Liddle syndrome in *Xenopus* oocytes could reproduce an abnormally high activity of ENaC, as postulated by G. Liddle in his seminal paper on pseudoaldosteronism. Indeed, deletion mutations or site-directed mutations targeting the PPP.Y sequence (PY motif) in β or γ ENaC, increase ENaC activity up to fourfold when expressed in *Xenopus* oocytes, due to an increase in the number and the activity of ENaC channels at the cell surface (Schild et al. 1995). These observations indicate that the PY motif in the cytoplasmic carboxy-terminus of ENaC critically modulates its activity at the cell surface (Firsov et al. 1996; Snyder et al. 1995).

The PY motif

During the last few years, there has been remarkable progress regarding our understanding of the molecular and cellular mechanisms underlying the regulatory role of the PY motif on ENaC activity. Using a two hybrid technique, Staub et al. first identified a binding partner for the β ENaC PY motif from a rat lung cDNA library (Staub et al. 1996). The partner found was a ubiquitin protein ligase called Nedd4, which is expressed not only in the lungs but in many other excitable and epithelial tissues. This discovery suggested that ubiquitination was a possible mechanism for ENaC regulation at the cell surface (Staub et al. 1997). The binding interaction between Nedd4 and the PY motif of ENaC involves conserved WW domains on Nedd4, in particular the third and the fourth WW domains of the protein (Kamynina et al. 2001b). The function of this Nedd4 protein is to catalyze the final steps in the ubiquitination cascade, in particular the attachment of ubiquitin moieties onto lysine residues in the N-terminus of α and γ ENaC (Staub et al. 1997). Then, the ubiquitinated ENaC channel at the cell surface undergoes endocytosis and likely lysosomal degradation.

Later, it was found in Na^+ transporting cells derived from the CCD that the likely candidate for interaction with ENaC is a homologue of Nedd4, called Nedd4-2 (Kamynina et al. 2001a). Coexpression of Nedd4-2 with ENaC in the *Xenopus* oocytes almost suppressed ENaC activity at the cell surface, likely because of Nedd4-2-induced increase in ENaC endocytosis and degradation. The contribution of Nedd4-2 in the pathogenesis of the Liddle syndrome was strongly suggested by the observation that a catalytically inactive Nedd4-2 obtained by site-directed mutagenesis reproduced in *Xenopus* oocytes a similar stimulatory effect of ENaC activity to the one obtained by mutations in the ENaC PY motif (Abriel et al. 1999). The prevailing idea underlying the molecular mechanisms causing the Liddle syndrome is that mutations in the PY motif of ENaC, by impairing the binding interaction with Nedd4-2, inhibit ENaC endocytosis and ENaC degradation in the lysosome, leading to an increased stability and therefore a longer half-life of ENaC activity at the cell surface.

WW domains, such as those found in Nedd4-2, recognize a large variety of proteins carrying proline-rich motifs. The specificity of the binding interaction between the WW domains of Nedd4-2 and the PY motif of ENaC has been addressed in functional experiments and in vitro binding assays. They show that the third WW (WW3) and the fourth WW (WW4) domains are required for the suppressor effect of Nedd4-2 on ENaC activity and that WW3 determines the high affinity of interaction with the ENaC PY motif (Kamynina et al. 2001b; Henry et al. 2003). The structure of the WW3 domain of Nedd4 with the β ENaC PY motif has been solved by NMR and reveals a three-stranded antiparallel β-sheet with a hydrophobic binding surface (Kanelis et al. 2001). The two prolines preceding the tyrosine of the PY motif form a polyproline type II helix and are accommodated in a groove of the WW3 domain. Substitution of the tyrosine residue 618 in β ENaC (Tyr-618 to Leu), as found in the Liddle syndrome, disrupts the hydrophobic interaction essential for recognition between the PY motif and Nedd4. Thus, the solved structure of the Nedd4 WW3 domain can fairly well explain the impact of ENaC mutations in patients with the Liddle syndrome on the binding interaction between Nedd4 and the PY motif of ENaC.

PY motifs and aldosterone response

Although Nedd4-2 controls ENaC activity at the cell surface, this protein is not a major aldosterone-regulated protein. This does not exclude the possibility that Nedd4-2 might be involved in the aldosterone cascade to ultimately regulate ENaC density at the cell surface. This hypothesis became quite attractive with the observation that the aldosterone-induced kinase SGK, like ENaC, contains in its amino-acid sequence a PY motif that could potentially be involved in binding interactions with the WW domains of Nedd4-2. It was found in the *Xenopus* oocyte expression system that Nedd4-2 interacts with SGK and that this interaction requires the PY motif of SGK (Debonneville et al. 2001; Snyder et al. 2002). The Nedd4-2-SGK interaction leads to the phosphorylation of Nedd4-2 by the kinase, which then reduces the Nedd4-2 ability to interact with ENaC. The consequence is an increase in activity of ENaC at the cell surface. These experiments point to a possible role of the PY motif in the ENaC response to aldosterone.

This novel signaling pathway could theoretically account for the stimulation of ENaC activity at the cell surface during the early response to aldosterone, since SGK is an early aldosterone-induced protein (Loffing et al. 2001b). According to this hypothesis, the increase in the number of active ENaC channels at the apical surface of ASDN cells elicited by aldosterone results from a prolonged half-life of activity at the cell surface due to inhibition of ENaC retrieval from the cell surface.

The aldosterone-SGK-Nedd4 pathway targets activate ENaC at the cell surface. However, as mentioned earlier, the number of ENaC molecules detected in the apical membrane is quite small in the absence of aldosterone. Thus, another aldosterone signaling pathway needs to be postulated to first promote ENaC translocation at the cell surface during the early phase of aldosterone stimulation. It is conceivable that in the CCD, aldosterone first translocates active ENaC at the cell surface, then inhibition of ENaC retrieval from the apical membrane by phosphorylation and inactivation of Nedd4-2 would further stimulate Na^+ absorption. However, it should be pointed out that this latter effect of aldosterone is expected to be efficient on Na^+ absorption only after 3–4 h, considering that the average half-life of the active ENaC at the cell surface is approximately 1 h (Hanwell et al. 2002).

The physiological and pathophysiological relevance of the aldosterone-SGK-Nedd4-2-ENaC signaling pathway has been addressed in an aldosterone-responding CCD cell line expressing ENaC Liddle mutants and in the mouse model of the Liddle syndrome (Auberson et al. 2003; Dahlmann et al. 2003). These experimental models allowed comparison of the effects of aldosterone on wild-type ENaC or ENaC lacking PY motifs in their C-termini. CCD cells expressing Liddle's ENaC mutants lacking the β and/or γ ENaC C-termini or substitution of Thyr in the β ENaC PY motif cells showed a 3–4-fold increase in baseline amiloride-sensitive Na^+ current. This is consistent with a retention of active ENaC at the cell surface, due to a prolonged half-life of ENaC activity at the cell surface (Auberson et al. 2003). In these cells, the rate of the increase in ENaC-mediated Na^+ transport during the first 3 h of aldosterone action was identical in both wild-type and mutant ENaC. Thus, the rate of insertion of ENaC in the apical membrane during the early response to aldosterone and vasopressin does not seem to be affected by PY motif mutations, nor by changes in the half-life of the channel activity at the cell surface.

Studies on isolated CCD cells from wild-type mice and homozygote Liddle mutant mice (L/L mice), have addressed the sensitivities to aldosterone during the late response phase (Dahlmann et al. 2003). Under a high salt diet and a low plasma aldosterone concentration, no amiloride-sensitive whole cell current was detected in principal cells, consistent with a low density of active ENaC in the apical membrane. When mice were placed under a low Na^+ diet for 6–8 days or were infused with aldosterone, ENaC-mediated Na^+ currents could be recorded in principal cells. Interestingly, the L/L mice exhibited considerably (up to 5 times) higher ENaC currents compared to wild-type mice. This exacerbated effect on aldosterone in L/L mice is consistent with a retention of long-acting ENaC mutants at the cell surface during the late response of aldosterone.

These studies on CCD cells expressing wild-type ENaC or mutant ENaC lacking the PY motif show that during the early phase (1–3 h) of aldosterone and vasopressin stimulation, no difference in Na^+ transport stimulation was observed, as if ENaC stimulation was independent of the integrity of the PY motif. During the late response of aldosterone (6–8 days), CCD cells expressing ENaC mutants lacking the PY motif exhibit a considerably higher Na^+ transport than wild-type cells, suggesting that mutations in the PY motif tend to potentiate the late aldosterone response. In other words, the Liddle mouse model tells us that the difference in Na^+ transport between wild-type and L/L mice is maximal under aldosterone stimulation.

In his original paper on pseudohypoaldosteronism, G.W. Liddle reported that when his patient was maintained on a constant diet of 30 mEq of sodium/day, and then given aldosterone, the urinary sodium excretion fell from approximately 30 mEq/day to less than 1 mEq/day (Liddle et al. 1963). This indicates that patients with ENaC mutations causing Liddle syndrome respond to aldosterone by increasing distal Na^+ reabsorption. We can postulate that in Liddle's patients the expansion of extracellular fluid volume likely starts to develop not only because of a higher baseline Na^+ absorption in the ASDN, but also because of an exacerbated and prolonged response of the ASDN to aldosterone stimulation. Later, as the extracellular volume expands, the plasma levels of aldosterone and renin drop, but volume expansion and hypertension persist. This suggests that the baseline Na^+ absorption in the ASDN of patients with the Liddle syndrome remains sufficiently high to maintain elevated blood pressure.

Until recently, all the reported cases of the Liddle syndrome were associated with mutations in the C-termini of β and γ ENaC. However, one hypertensive proband from a Finnish family with the characteristic clinical features of the Liddle syndrome was shown to harbor a gain-of-function mutation (γN530S) in the extracellular domain of γ ENaC (Hiltunen et al. 2002). The Asn 530-to-Ser substitution is located in a stretch of highly conserved amino acids preceding the second transmembrane segment, and was found to enhance ENaC activity in the *Xenopus* expression system. The molecular basis for the ENaC γN530S-induced gain of function is not perfectly clear. A corresponding amino acids' substitution in the *C. elegans* ortholog of ENaC (MEC4, DEG-1 genes) causes cell swelling and neuronal degeneration that were interpreted as a channel hyperactivity (Driscoll and Chalfie 1991). Mutations at neighboring positions in α and β ENaC were investigated by amino acid substitutions with Cys and subsequent modifications by sulfhydryl reagents (Kellenberger et al. 2002; Snyder et al. 2000). Covalent modifications of cysteine residues at a position adjacent to γAsn530 induce conformational changes that lock the channel in the open configuration, suggesting that the region encompassing γAsn530 is part of a gate that controls ENaC openings and closures. This extracellular gate is located close to or within the external entrance of the channel and likely controls the access of the ENaC pore by permeant ions. This region, located in the exon 13 of the genes coding the α, β, and γ ENaC subunits, represents a target for mutations with a likely important impact on channel function.

Loss-of-function mutations

The systemic type 1 pseudohypoaldosteronism is caused by homozygous mutations in any one of the three ENaC subunits. The type of mutations and their functional impact on ENaC have been extensively reviewed recently (Bonny and Rossier 2002). The majority of mutations associated with PHA-1 are frameshift or nonsense mutations leading to truncated or completely abnormal proteins, which, not surprisingly, are nonfunctional. The missense mutations associated with PHA-1 were more informative for our understanding of ENaC structure and function.

A homozygous G37S missense mutation in β ENaC, which is located in an absolutely conserved HG motif in the N-terminus of all the members of the ENaC/DEG channel family, is associated with a mild form of PHA-1. This HG motif is approximately 15 amino acid residues upstream of the TM1. According to current membrane topology models of ENaC and degenerins, this motif is facing the cytoplasmic side of the membrane. The functional impact of the βG37S mutation was a 50% decrease in ENaC activity, consistent with the mild phenotype (Grunder et al. 1997). The corresponding mutation in the α ENaC subunit, which introduces two G-to-S substitutions in the channel, resulted in a 95% decrease in ENaC activity, which was essentially due to a decrease in the open probability. Thus, the G37S mutation responsible for a mild form of PHA-1 identified a domain in the cytoplasmic N-terminus of ENaC that controls channel gating. The physiological role of this "gating domain" remains to be established.

An αC133Y mutation that substitutes a Cys residue in the first cystein-rich domain of the extracellular loop was found to be associated with PHA-1. The presence of cystein-rich domains in the large extracellular loop is a characteristic of all the members of the ENaC/degenerin channel family. Evidence suggests that these extracellular cysteines are involved in disulfide bonds, but so far no specific channel function has been found to be linked with these conserved cysteines (Firsov et al. 1999). The Cys-to-Tyr substitution in α ENaC associated with PHA-1 may impair the correct folding of the extracellular loop, leading to abnormal routing of ENaC to the cell surface.

Another missense mutation causing PHA-1 was found at the 5′ start of the second transmembrane domain and substitutes a conserved Ser to Leu at position 562 of α ENaC (Schaedel et al. 1999). The Ser562 residue is lining the channel pore at its narrowest part and is therefore important for the ionic selectivity of the channel, allowing small ions like Na^+ or Li^+ to pass through the channel and excluding larger ions like K^+ or Cs^+ (Kellenberger et al. 1999a). Indeed, Ser 562 substitutions with amino acids of relatively similar sizes make ENaC permeant to K^+, Cs^+, or even large organic cations (Kellenberger et al. 1999b). With less conserved substitutions of Ser562 with aromatic residues, for instance, or in the case of S562L mutant, the channel is nonconducting, likely because of important steric changes in these critical regions of the channel pore that become incompatible with ion permeation and/or ENaC subunit assembly. It should be pointed out that the S562 residue that is part of the channel selectivity filter is located close to the amiloride binding site in the external entrance of the channel pore.

Conclusion

Genetic analysis of blood pressure variations in patients with electrolyte disorders, such as Liddle syndrome or PHA-1, has demonstrated that ENaC plays a critical role in the maintenance of Na^+ balance, extracellular fluid volume, and blood pressure. The studies of mutations causing ENaC loss of function and PHA-1 have identified important functional domains that are involved in ion permeation, and channel gating. The gain-of-function mutations associated with Liddle syndrome target proline-rich PY motifs in the cytoplasmic carboxy-terminus of ENaC that controls channel expression at the cell surface. The functional consequences of these mutations are retention of active channels at the cell surface and an exacerbated response to aldosterone, leading to abnormally high Na^+ absorption in the ASDN. Interestingly, the capacity of ENaC to mediate Na^+ absorption along the ASDN is not uniform, but seems to predominate in the CNT. In order to precisely understand how the ASDN controls Na^+ balance, it is now essential to relate our knowledge of ENaC function and expression in the ASDN with the emerging understanding of the signaling pathways of key hormones, such as aldosterone or vasopressin, that regulate ENaC activity in the ASDN.

References

Abriel H, Loffing J, Rebhun JF, Pratt JH, Schild L, Horisberger JD, Rotin D, Staub O (1999) Defective regulation of the ENaC by Nedd4 in Liddle's syndrome. J Clin Invest 103:667–673

Almeida AJ, Burg MB (1982) Sodium transport in the rabbit connecting tubule. Am JPhysiol 243:F330–F334

Alvarez de la Rosa D, Zhang P, Naray-Fejes-Toth A, Fejes-Toth G, Canessa CM (1999) The serum and glucocorticoid kinase sgk increases the abundance of epithelial sodium channels in the plasma membrane of Xenopus oocytes. J Biol Chem 274:37834–37839

Auberson M, Hoffmann-Pochon N, Vandewalle A, Kellenberger S, Schild L (2003) ENaC mutants causing Liddle's syndrome retain ability to respond to aldosterone and vasopressin. Am J Physiol Renal Physiol 285:F459–F471

Barker PM, Nguyen MS, Gatzy JT, Grubb B, Norman H, Hummler E, Rossier B, Boucher RC, Koller B (1998) Role of γ-ENaC subunit in lung liquid clearance and electrolyte balance in newborn mice. J Clin Invest 102:1634–1640

Bonny O, Rossier BC (2002) Disturbances of Na/K balance: pseudohypoaldosteronism revisited. J Am Soc Nephrol 13:2399–2414

Boucher RC (2004) New concepts of the pathogenesis of cystic fibrosis lung disease. Eur Respir J 23:146–158

Burch LH, Talbot CR, Knowles MR, Canessa CM, Rossier BC, Boucher RC (1995) Relative expression of the human ENaC subunits in normal and cystic fibrosis airways. Am J Physiol 269:C511–C518

Canessa CM, Horisberger JD, Rossier BC (1993) Epithelial sodium channel related to proteins involved in neurodegeneration. Nature 361:467–470

Canessa CM, Schild L, Buell G, Thorens B, Gautschi I, Horisberger JD, Rossier BC (1994) Amiloride-sensitive ENaC is made of three homologous subunits. Nature 367:463–467

Chang SS, Gründer S, Hanukoglu A, Rösler A, Mathew PM, Hanukoglu I, Schild L, Lu Y, Shimkets RA, Nelson-Williams C, Rossier BC, Lifton RP (1996) Mutations in subunits of the epithelial sodium channel cause salt wasting with hyperkalaemic acidosis, pseudohypoaldosteronism type 1. Nat Genet 12:248–253

Chen SY, Bhargava A, Mastroberardino L, Meijer OC, Wang J, Buse P, Firestone GL, Verrey F, Pearce D (1999) Epithelial sodium channel regulated by aldosterone-induced protein sgk. Proc Natl Acad Sci 96:2514–2519

Coscoy S, Lingueglia E, Lazdunski M, Barbry P (1998) The phe-met-arg-phe-amide-activated sodium channel is a tetramer. J Biol Chem 273:8317–8322

Costanzo LS (1984) Comparison of calcium and sodium transport in early and late rat distal tubules: effect of amiloride. Am J Physiol 246:F937–F945

Dahlmann A, Pradervand SP, Hummler E, Rossier BC, Frindt G, Palmer LG (2003) Mineralocorticoid regulation of ENaC is maintained in a mouse model of Liddle's syndrome. Am J Physiol Renal Physiol 285:F310–F318

Debonneville C, Flores SY, Kamynina E, Plant PJ, Tauxe C, Thomas MA, Munster C, Chraibi A, Pratt JH, Horisberger JD, Pearce D, Loffing J, Staub O (2001) Phosphorylation of Nedd4-2 by Sgk1 regulates ENaC cell surface expression. EMBO J 20:7052–7059

Dijkink L, Hartog A, Van Os CH, Bindels RJM (2002) The epithelial sodium channel ENaC is intracellularly located as a tetramer. Pflugers Arch-Eur J Physiol 444:549–555

Dinudom A, Young JA, Cook DI (1993) Amiloride-sensitive Na+ current in the granular duct cells of mouse mandibular glands. Pflugers Arch. 423:164–166

Driscoll M, Chalfie M (1991) The mec-4 gene is a member of a family of Caenorhabditis elegans genes that can mutate to induce neuronal degeneration. Nature 349:588–593

Duc C, Farman N, Canessa CM, Bonvalet JP, Rossier BC (1994) Cell-specific expression of epithelial sodium channel α, β, and γ subunits in aldosterone-responsive epithelia from the rat: Localization by in situ hybridization and immunocytochemistry. J Cell Biol 127:1907–1921

Epple HJ, Amasheh S, Mankertz J, Goltz M, Schulzke JD, Fromm M (2000) Early aldosterone effect in distal colon by transcriptional regulation of ENaC subunits. Am J Physiol Gastrointest Liver Physiol 278:G718–G724

Farman N, Talbot CR, Boucher R, Fay M, Canessa C, Rossier B, Bonvalet JP (1997) Noncoordinated expression of α, β, and g subunit mRNAs of epithelial Na^+ channel along rat respiratory tract. Am J Physiol Cell Physiol 272:C131–C141

Firsov D, Schild L, Gautschi I, Mérillat AM, Schneeberger E, Rossier BC (1996) Cell surface expression of the epithelial Na channel and a mutant causing Liddle syndrome: A quantitative approach. Proc Natl Acad Sci 93:15370–15375

Firsov D, Gautschi I, Merillat AM, Rossier BC, Schild L (1998) The heterotetrameric architecture of the epithelial sodium channel ENaC. EMBO J 17:344–352

Firsov D, Robert-Nicoud M, Gruender S, Schild L, Rossier BC (1999) Mutational analysis of cysteine-rich domains of the epithelium sodium channel ENaC—identification of cysteines essential for channel expression at the cell surface. J Biol Chem 274:2743–2749

Frindt G, Palmer LG (2003) Na channels in the rat connecting tubule. Am J Physiol Renal Physiol

Frindt G, Masilamani S, Knepper MA, Palmer LG (2001) Activation of epithelial Na channels during short-term Na deprivation. Am J Physiol 280:F112–F118

Frindt G, McNair T, Dahlmann A, Jacobs-Palme E, Palmer LG (2002) Epithelial Na channels and short-term renal response to salt deprivation. Am J Physiol Renal Physiol 283:F717–F726

Geller DS, Rodriguezsoriano J, Boado AV, Schifter S, Bayer M, Chang SS, Lifton RP (1998) Mutations in the mineralocorticoid receptor gene cause autosomal dominant pseudohypoaldosteronism type 1. Nat Genet 19:279–281

Geller DS, Farhi A, Pinkerton N, Fradley M, Moritz M, Spitzer A, Meinke G, Tsai FTF, Sigler PB, Lifton RP (2000) Activating mineralocorticoid receptor mutation in hypertension exacerbated by pregnancy. Science 289:119–123

Grantham JJ, Burg MB, Orloff J (1970) The nature of transtubular sodium and potassium transport in isolated rabbit renal cortical collecting tubules. J Clin Invest 49:1815–1826

Grunder S, Firsov D, Chang SS, Jaeger NF, Gautschi I, Schild L, Lifton RP, Rossier BC (1997) A mutation causing pseudohypoaldosteronism type 1 identifies a conserved glycine that is involved in the gating of the epithelial sodium channel. EMBO J 16:899–907

Guyton AC (1992) Kidneys and fluids in pressure regulation. Small volume but large pressure changes. Hypertension 19:12–18

Hager H, Kwon TH, Vinnikova AK, Masilamani S, Brooks HL, Frokiaer J, Knepper MA, Nielsen S (2001) Immunocytochemical and immunoelectron microscopic localization of α-, β-, and γ-ENaC in rat kidney. Am J Physiol Renal Physiol 280:F1093–F1106

Hanukoglu A (1991) Type I pseudohypoaldosteronism includes two clinically and genetically distinct entities with either renal or multiple target organ defects. J Clin Endocrinol Metab 73:936–944

Hanwell D, Ishikawa T, Saleki R, Rotin D (2002) Trafficking and cell surface stability of the epithelial Na+ channel expressed in epithelial Madin-Darby canine kidney cells. J Biol Chem 277:9772–9779

Hawk CT, Li L, Schafer JA (1996) AVP and aldosterone at physiological concentrations have synergistic effects on Na+ transport in rat CCD. Kidney Int Suppl 57:S35–41

Henry PC, Kanelis V, O'Brien MC, Kim B, Gautschi I, Forman-Kay J, Schild L, Rotin D (2003) Affinity and specificity of interactions between Nedd4 isoforms and the epithelial Na+ channel. J Biol Chem 278:20019–20028

Hiltunen TP, Hannila-Handelberg T, Petajaniemi N, Kantola I, Tikkanen I, Virtamo J, Gautschi I, Schild L, Kontula K (2002) Liddle's syndrome associated with a point mutation in the extracellular domain of the epithelial sodium channel gamma subunit. J Hypertens 20:2383–2390

Hummler E, Barker P, Gatzy J, Beermann F, Verdumo C, Schmidt A, Boucher RC, Rossier BC (1996) Early death due to defective neonatal lung liquid clearance in α ENaC-deficient mice. Nat Genet 12:325–328

Kamynina E, Debonneville C, Bens M, Vandewalle A, Staub O (2001a) A novel mouse Nedd4 protein suppresses the activity of the epithelial Na+ channel. FASEB J 15:204–214

Kamynina E, Tauxe C, Staub O (2001b) Distinct characteristics of two human Nedd4 proteins with respect to epithelial Na+ channel regulation. Am J Physiol Renal Physiol 281:F469–F477

Kanelis V, Rotin D, Forman-Kay JD (2001) Solution structure of a Nedd4 WW domain-ENaC peptide complex. Nat Struct Biol 8:407–412

Kellenberger S, Schild L (2002) Epithelial sodium channel/degenerin family of ion channels: a variety of functions for a shared structure. Physiol Rev 82:735–767

Kellenberger S, Gautschi I, Schild L (1999a) A single point mutation in the pore region of the epithelial Na+ channel changes ion selectivity by modifying molecular sieving. Proc Natl Acad Sci 96:4170–4175

Kellenberger S, Hoffmann-Pochon N, Gautschi I, Schneeberger E, Schild L (1999b) On the molecular basis of ion permeation in the epithelial Na+ channel. J Gen Physiol 114:13–30

Kellenberger S, Gautschi I, Schild L (2002) An external site controls closing of the epithelial Na+ channel ENaC. J Physiol London 543:413–424

Koefoed-Johnson V, Ussing HH (1958) The nature of frog skin potential. Acta Physiol Scand 42:298–308

Kosari F, Sheng SH, Li JQ, Mak DD, Foskett JK, Kleyman TR (1998) Subunit stoichiometry of the epithelial sodium channel. J Biol Chem 273:13469–13474

Liddle GW, Bledsoe T, Coppage WS (1963) A familial renal disorder simulating primary aldosteronism but with negligible aldosterone secretion. Trans Assoc Am Physicians 76:199–213

Lifton RP (1996) Molecular genetics of human blood pressure variation. Science 272:676–680

Lifton RP, Gharavi AG, Geller DS (2001) Molecular mechanisms of human hypertension (review). Cell 104:545–556

Lindemann B (1996) Taste reception. Physiol Rev 76:718–766

Lingueglia E, Voilley N, Waldmann R, Lazdunski M, Barbry P (1993) Expression cloning of an epithelial amiloride-sensitive Na^+ channel. FEBS Lett 318:95–99

Loffing J, Kaissling B (2003) Sodium and calcium transport pathways along the mammalian distal nephron: from rabbit to human. Am J Physiol Renal Physiol 284:F628–F643

Loffing J, Loffing-Cueni D, Macher A, Hebert SC, Olson B, Knepper MA, Rossier BC, Kaissling B (2000a) Localization of epithelial sodium channel and aquaporin-2 in rabbit kidney cortex. Am J Physiol Renal Physiol 278:F530–F539

Loffing J, Pietri L, Aregger F, Bloch-Faure M, Ziegler U, Meneton P, Rossier BC, Kaissling B (2000b) Differential subcellular localization of ENaC subunits in mouse kidney in response to high- and low-Na diets. Am J Physiol Renal Physiol 279:F252–F258

Loffing J, Loffing-Cueni D, Valderrabano V, Klausli L, Hebert SC, Rossier BC, Hoenderop JGJ, Bindels RJM, Kaissling B (2001a) Distribution of transcellular calcium and sodium transport pathways along mouse distal nephron. Am J Physiol Renal Physiol 281:F1021–F1027

Loffing J, Zecevic M, Feraille E, Kaissling B, Asher C, Rossier BC, Firestone GL, Pearce D, Verrey F (2001b) Aldosterone induces rapid apical translocation of ENaC in early portion of renal collecting system: possible role of SGK. Am J Physiol Renal Physiol 280:F675–F682

May A, Puoti A, Gaeggeler HP, Horisberger JD, Rossier BC (1997) Early effect of aldosterone on the rate of synthesis of the epithelial sodium channel α subunit in A6 renal cells. J Am Soc Nephrol 8:1813–1822

McDonald FJ, Yang BL, Hrstka RF, Drummond HA, Tarr DE, McCray PB, Stokes JB, Welsh MJ, Williamson RA (1999) Disruption of the β subunit of the epithelial Na^+ channel in mice: hyperkalemia and neonatal death associated with a pseudohypoaldosteronism phenotype. Proc Natl Acad Sci 96:1727–1731

Morris RG, Schafer JA (2002) cAMP increases density of ENaC subunits in the apical membrane of MDCK cells in direct proportion to amiloride-sensitive Na+ transport. J Gen Physiol 120:71–85

Naray FT, Canessa C, Cleaveland ES, Aldrich G, Fejes-Toth G (1999) SgK is an aldosterone-induced kinase in the renal collecting duct—effects on epithelial Na+ channels. J Biol Chem 274:16973–16978

Nielsen S, Chou CL, Marples D, Christensen EI, Kishore BK, Knepper MA (1995) Vasopressin increases water permeability of kidney collecting duct by inducing translocation of aquaporin-CD water channels to plasma membrane. Proc Natl Acad Sci 92:1013–1017

Palmer LG, Frindt G (1986) Amiloride-sensitive Na channels from the apical membrane of the rat cortical collecting tubule. Proc Natl Acad Sci 83:2727–2770

Palmer LG, Edelman IS, Lindemann B (1980) Current-voltage analysis of apical sodium transport in toad urinary bladder: effects of inhibitors of transport and metabolism. J Membr Biol 57:59–71

Pradervand SP, Barker PM, Wang Q, Ernst SA, Beermann F, Grubb BR, Burnier M, Schmidt A, Bindels RM, Gatzy JT, Rossier BC, Hummler E (1999) Salt restriction induces pseudohypoaldosteronism type 1 in mice expressing low levels of the β-subunit of the amiloride-sensitive epithelial sodium channel. Proc Natl Acad Sci 96:1732–1737

Robert-Nicoud M, Flahaut M, Elalouf JM, Nicod M, Salinas M, Bens M, Doucet A, Wincker P, Artiguenave F, Horisberger JD, Vandewalle A, Rossier BC, Firsov D (2001) Transcriptome of a mouse kidney cortical collecting duct cell line: effects of aldosterone and vasopressin. Proc Natl Acad Sci 98:2712–2716

Rossier BC, Pradervand SP, Schild L, Hummler E (2002) Epithelial sodium channel and the control of sodium balance: interaction between genetic and environmental factors. Annu Rev Physiol 64:877–897

Rubera I, Loffing J, Palmer LG, Frindt G, Fowler-Jaeger N, Sauter D, Carroll T, McMahon A, Hummler E, Rossier BC (2003) Collecting duct-specific gene inactivation of αENaC in the mouse kidney does not impair sodium and potassium balance. J Clin Invest 112:554–565

Schaedel C, Marthinsen L, Kristoffersson AC, Kornfalt R, Nilsson KO, Orlenius B, Holmberg L (1999) Lung symptoms in pseudohypoaldosteronism type 1 are associated with deficiency of the α-subunit of the epithelial sodium channel. J Pediatr 135:739–745

Schild L, Canessa CM, Shimkets RA, Gautschi I, Lifton RP, Rossier BC (1995) A mutation in the epithelial sodium channel causing Liddle disease increases channel activity in the Xenopus laevis oocyte expression system. Proc Natl Acad Sci 92:5699–5703

Shimkets RA, Warnock DG, Bositis CM, Nelson-Williams C, Hansson JH, Schambelan M, Gill JR Jr, Ulick S, Milora RV, Findling JW, Canessa CM, Rossier BC, Lifton RP (1994) Liddle's syndrome: heritable human hypertension caused by mutations in the β subunit of the epithelial sodium channel. Cell 79:407–414

Snyder PM, Price MP, McDonald FJ, Adams CM, Volk KA, Zeiher BG, Stokes JB, Welsh MJ (1995) Mechanism by which Liddle's syndrome mutations increase activity of a human epithelial Na^+ channel. Cell 83:969–978

Snyder PM, Bucher DB, Olson DR (2000) Gating induces a conformational change in the outer vestibule of ENaC. J Gen Physiol 116:781–790

Snyder PM, Olson DR, Thomas BC (2002) Serum and glucocorticoid-regulated kinase modulates Nedd4-2-mediated inhibition of the epithelial Na+ channel. J Biol Chem 277:5–8

Staub O, Gautschi I, Ishikawa T, Breitschopf K, Ciechanover A, Schild L, Rotin D (1997) Regulation of stability and function of the epithelial Na channel ENaC by ubiquitination. EMBO J 16:6325–6336

Staub O, Dho S, Henry PC, Correa J, Ishikawa T, McGlade J, Rotin D (1996) WW domains of Nedd4 bind to the proline-rich PY motifs in the epithelial Na^+ channel deleted in Liddle's syndrome. EMBO J 15:2371–2380

Talbot CL, Bosworth DG, Briley EL, Fenstermacher DA, Boucher RC, Gabriel SE, Barker PM (1999) Quantitation and localization of ENaC subunit expression in fetal, newborn, and adult mouse lung. Am J Respir Cell Mol Biol 20:398–406

van Balkom BW, Graat MP, van Raak M, Hofman E, van der Sluijs SP, Deen PM (2004) Role of cytoplasmic termini in sorting and shuttling of the aquaporin-2 water channel. Am J Physiol Cell Physiol 286:C372–C379

Verrey F, Hummler E, Schild L, Rossier BC (2001) Control of Na^+ transport by aldosterone. In: Seldin DW, Giebisch G (eds) The kidney. Lippincott, Philadelphia, pp 1441–1472

Instructions for authors

1 Legal requirements

The author(s) guarantee(s) that the manuscript will not be published elsewhere in any language without the consent of the copyright holders, that the rights of third parties will not be violated, and that the publisher will not be held legally responsible should there be any claims for compensation.

Authors wishing to include figures or text passages that have already been published elsewhere are required to obtain permission from the copyright holder(s) and to include evidence that such permission has been granted when submitting their papers. Any material received without such evidence will be assumed to originate from the authors.

Manuscripts must be accompanied by the "Copyright Transfer Statement".

Please include at the end of the acknowledgements a declaration that the experiments comply with the current laws of the country in which they were performed.

2 Editorial procedure

Manuscripts should be submitted in English, together with one set of illustrations and a complete pdf file, to the editor in charge.

The author is responsible for the accuracy of the references.

3 Manuscript preparation

To help you prepare your manuscript, Springer offers a template that can be used with Winword 7 (Windows 95), Winword 6 and Word for Macintosh.

For details see point 4.

All manuscripts are subject to copy editing.

■ **Title page**
- The name(s) of the author(s)
- A concise and informative title
- The affiliation(s) and address(es) of the author(s)
- The e-mail address, telephone and fax numbers of the communicating author

■ **Abstract.** Each paper must be preceded by an abstract presenting the most important results and conclusions.

■ **Abbreviations** should be defined at first mention in the abstract and again in the main body of the text and used consistently thereafter

A list of **symbols** should follow the abstract if such a list is needed. Symbols must be written clearly. The international system of units (SI units) should be used. The numbering of chapters should be in decimal form.

Footnotes on the title page are not given reference symbols. Footnotes to the text are numbered consecutively; those to tables should be indicated by superscript lower-case letters (or asterisks for significance values and other statistical data).

Acknowledgements. These should be as brief as possible. Any grant that requires acknowledgement should be mentioned. The names of funding organizations should be written in full.

Funding. Authors are expected to disclose any commercial or other associations that might pose a conflict of interest in connection with submitted material. All funding sources supporting the work and institutional or corporate affiliations of the authors should be acknowledged.

■ **References**

The list of References should only include works that are cited in the text and that have been published or accepted for publication. Personal communications should only be mentioned in the text.

In the text, references should be cited by author and year (e.g. Hammer 1994; Hammer and Sjöqvist 1995; Hammer et al. 1993) and listed in alphabetical order in the reference list.

Examples:

Monographs:
Snider T, Grand L (1982) Air pollution by nitrogen oxides. Elsevier, Amsterdam

Anthologies and proceedings:
Noller C, Smith VR (1997) Ultraviolet selection pressure on earliest organisms. In: Kingston H, Fulling CP (eds) Natual environment background analysis. Oxford University Press, Oxford, pp 211–219

Journals:
Meltzoff AN, Moore MK (1977) Imitation of facial and manual gestures by human neonates. Science 198:75–78

If available the Digital Object Identifier (DOI) of the cited literature should be added at the end of the reference in question.

■ **Illustrations and Tables**

All figures (photographs, graphs or diagrams) and tables should be cited in the text, and each numbered consecutively throughout. Figure parts should be identified by lower-case roman letters. The placement of figures and tables should be indicated in the left margin. For submission of figures in electronic form see below

Line drawings. Please submit good-quality prints. The inscriptions should be clearly legible.

Half-tone illustrations (black and white and color). Please submit well-contrasted photographic prints with the top indicated on the back. Magnification should be indicated by scale bars.

Figure legends must be brief, self-sufficient explanations of the illustrations. The legends should be placed at the end of the text.

Tables should have a title and a legend explaining any abbreviation used in that table. Footnotes to tables should be indicated by superscript lower-case letters (or asterisks) for significance values and other statistical data.

4 Electronic submission of final version

Please send only the final version of the article, as accepted by the editors.

Preparing your manuscript

The template is available:

→ via ftp:
Address: ftp.springer.de/
User ID: ftp
Password: your own e-mail address
– Directory: /pub/Word/journals
– File names: either sv-journ.zip or sv-journ.doc and sv-journ.dot

→ via browser
– http://www.springer.de/author/index.html

The zip file should be sent uuencoded.

Layout guidelines

1. Use a normal, plain font (e.g., Times Roman) for text.
   Other style options:
   – for textual emphasis use italic types.
   – for special purposes, such as for mathematical vectors, use boldface type.
2. Use the automatic page numbering function to number the pages.
3. Do not use field functions.
4. For indents use tab stops or other commands, not the space bar.
5. Use the table functions of your word processing program, not spreadsheets, to make tables.
6. Use the equation editor of your word processing program or MathType for equations.
7. Place any figure legends or tables at the end of the manuscript.
8. Submit all figures as separate files and do not integrate them within the text.

Data formats

Save your file in two different formats:

1. RTF (Rich Text Format) or Word compatible Word 95/97
2. pdf (a single pdf file including text, tables and figures)

Illustrations

The preferred figure formats are EPS for vector graphics exported from a drawing program and TIFF for halftone illustrations. EPS files must always contain a preview in TIFF of the figure. The file name (one file for each figure) should include the figure number. Figure legends should be included in the text and not in the figure file.

Scan resolution: Scanned line drawings should be digitized with a minimum resolution of 800 dpi relative to the final figure size. For digital halftones, 300 dpi is usually sufficient.

Color illustrations: Store color illustrations as RGB (8 bits per channel) in TIFF format.

General information on data delivery

Please send us a zip file (text and illustrations in separate files) either:

→ Via ftp.springer.de
(to our ftp.server; log-in "anonymous"; password: your e-mail address; further information in the readme file on the server)

→ By e-mail
(only suitable for small volumes of data)
→ or on any of the following media:
– On a diskette [you may use .tar, .zip, .gzip (.gz), .sit, and compress (.Z)]
– On a ZIP cartridge
– On a CD-ROM

Please always supply the following information with your data: journal title, operating system, word processing program, drawing program, image processing program, compression program.

The file name should be memorable (e.g., author name), have no more than 8 characters, and include no accents or special symbols. Use only the extensions that the program assigns automatically.

5 Proofreading

Authors should make their proof corrections on a printout of the pdf file supplied, checking that the text is complete and that all figures and tables are included. After online publication, further changes can only be made in the form of an Erratum, which will be hyperlinked to the article. The author is entitled to formal corrections only. Substantial changes in content, e.g. new results, corrected values, title and authorship are not allowed without the approval of the editor in charge. In such a case please contact the Editor in charge before returning the proofs to the publisher.

6 Offprints, Free copy

You are entitled to receive a pdf file of your article for your own personal use. Orders for offprints can be placed by returning the order form with the corrected proofs. One complimentary copy of the issue in which your article appears is supplied.

MIX
Papier aus verantwortungsvollen Quellen
Paper from responsible sources
FSC® C105338

If you have any concerns about our products,
you can contact us on
ProductSafety@springernature.com

In case Publisher is established outside the EU,
the EU authorized representative is:
Springer Nature Customer Service Center GmbH
Europaplatz 3, 69115 Heidelberg, Germany

Printed by Libri Plureos GmbH
in Hamburg, Germany